PROTEIN TARGETING
AND SECRETION

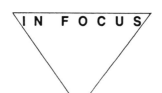

Titles published in the series:

*Antigen-presenting Cells

*Complement

DNA Replication

Enzyme Kinetics

Gene Structure and Transcription

Genetic Engineering

*Immune Recognition

*B Lymphocytes

*Lymphokines

Membrane Structure and Function

Molecular Basis of Inherited Disease

Protein Engineering

Protein Targeting and Secretion

Regulation of Enzyme Activity

*Published in association with the British Society for Immunology.

Series editors

David Rickwood
Department of Biology, University of Essex, Wivenhoe Park,
Colchester, Essex CO4 3SQ, UK

David Male
Institute of Psychiatry, De Crespigny Park, Denmark Hill,
London SE5 8AF, UK

PROTEIN TARGETING AND SECRETION

Brian M.Austen and
Olwyn M.R.Westwood

Department of Surgery, St George's Medical School,
London SW17 0RE, UK

⬡IRL PRESS
——at——
OXFORD UNIVERSITY PRESS

Oxford University Press
Walton Street, Oxford OX2 6DP

Oxford is a trade mark of Oxford University Press

Published in the United States
by Oxford University Press, New York

© Oxford University Press 1991

*All rights reserved. No part of this publication may be reproduced, stored in a
retrieval system, or transmitted, in any form or by any means, electronic,
mechanical, photocopying, recording, or otherwise, without the prior permission
of Oxford University Press.*

*This book is sold subject to the condition that it shall not, by way of trade or
otherwise, be lent, re-sold, hired out, or otherwise circulated without the
publisher's prior consent in any form of binding or cover other than that in
which it is published and without a similar condition including this condition
being imposed on the subsequent purchaser.*

British Library Cataloguing in Publication Data
Austen, Brian M.
Protein targeting and secretion.
1. Organisms. Cells. Proteins. Processing & turnover
I. Title   II. Westwood, Olwyn M.R.   III. Series 574.8761
ISBN 0-19-963217-0

Library of Congress Cataloging in Publication Data
Austen, Brian M.
Protein targeting and secretion/Brian M. Austen and
Olwyn M.R. Westwood
(In focus)
Includes bibliographical references.
Includes index.
1. Proteins – Physiological transport.   2. Proteins – Secretion.
I. Westwood, Olwyn M.R.   II. Title.   III. Series: In focus
(Oxford, England)
[DNLM: 1. Biological Transport.   2. Cell Communication.   3. Protein
Binding.   4. Proteins – secretion.   5. Translocation (Genetics)
QU 55 A933p]
QP551.A97 1991   612'. 01575 – dc20   90 – 14346.
ISBN 0-19-963217-0 (pbk.)

Typeset and printed by Information Press Ltd, Oxford, England.

Preface

Although an electron micrograph gives a static picture of the substructure of a living cell, in reality there is considerable change and movement. Proteins that are synthesized in the cytoplasm rapidly move to their final destinations within the cell, while other proteins are secreted or taken into the cell from the environment. In the course of migration, a protein's structure changes markedly from what was, at synthesis, a flexible linear chain emerging from a ribosome to one that has folded to a defined stereochemistry. In addition, other substances may become covalently linked to the protein, or portions of a protein's structure may be proteolytically removed. How proteins find their way among the various membranes of subcellular organelles depends on the decoding of signals found within their structure by receptors or enzymes.

Our aim in this book is not only to give an outline of the mechanisms of protein targeting, but also to provide some insight, albeit highly selective, into the experimental techniques that have been used. We are indebted to a number of researchers who have freely donated photographs of their results.

Brian M. Austen
Olwyn M.R. Westwood

Contents

Abbreviations

ACTH	adrenocorticotrophic hormone
BiP	heavy chain binding protein
CD	circular dichroism
DHFR	dihydrofolate reductase
ER	endoplasmic reticulum
ERD2	endoplasmic reticulum retention defective (gene 2)
GAP	GTP-ase activating protein
GroEL(S)	genes involved in the maturation of λ-phage
HDEL	His – Asp – Glu – Leu
hsp	heat-shock protein
IMS	intermembrane space
KDEL	Lys – Asp – Glu – Leu
MBP	maltose binding protein
2D-NMR	2-dimensional nuclear magnetic resonance
NSF	N-ethyl maleimide-sensitive fusion protein
p21^{ras}	21 kDa protein encoded by murine sarcoma virus oncogene
p60^{src}	60 kDa protein encoded by rous sarcoma virus oncogene
PDI	protein disulphide isomerase
PLC	PI specific phospholipase C
rab	genes from rat brain homologous to ras
ras	oncogene from mouse sarcoma virus
Rubisco	ribulose-1,5-bisphosphate carboxylase-oxygenase
SDS-PAGE	polyacrylamide gel electrophoresis in sodium dodecyl sulphate
SNAP	soluble NSF attachment protein
SRP	signal recognition particle
VSV-G	vesicular stomatis virus G-protein
VPS	vacuolar protein sorting genes

Amino acid	One-letter symbol	Three-letter symbol
Alanine	A	Ala
Arginine	R	Arg
Asparagine	N	Asn
Aspartic acid	D	Asp
Cysteine	C	Cys
Glutamine	Q	Glu
Glutamic acid	E	Gln
Glycine	G	Gly
Histidine	H	His
Isoleucine	I	Ile
Leucine	L	Leu
Lysine	K	Lys
Methionine	M	Met
Phenylalanine	F	Phe
Proline	P	Pro
Serine	S	Ser
Threonine	T	Thr
Tryptophan	W	Trp
Tyrosine	Y	Tyr
Valine	V	Val
Unknown or 'other'	X	

1

General principles

1. Subcellular organization

Prokaryotes are single-cell organisms that do not contain subcellular membrane-bound structures, whereas the eukaryotic cell is divided into distinct subcellular compartments or organelles. Up to 95 per cent of the eukaryotic cell's membranes is in intracellular structures. Compartments include the nucleus, endoplasmic reticulum (ER), peroxisome, and mitochondrion. Each compartment is enclosed by one or more membranes and mitochondria and chloroplasts are subdivided further by membranes into subcompartments. There is considerable traffic of proteins moving from one compartment or subcompartment to another, from the cytoplasm to intracellular compartments or out of the cell, and from the extracellular environment into the cell's interior. In the course of movement many proteins, or parts of proteins, which are essentially polar macromolecules, cross from one side of the hydrophobic barrier of a phospholipid bilayer to the other.

2. Targeting sequences and patches

Many proteins are targeted to their destinations by small parts of their structure (*Figure 1.1*) (1). In some instances, these sequences are cleaved off upon arrival of the protein by a peptidase that is situated in the destination compartment. Targeting sequences that occur at the amino-terminal end of the protein include the hydrophobic sequences known as signal sequences that target proteins through the ER membrane, or bacterial cytoplasmic membrane. Amphiphilic sequences at the amino-terminus of precursors that guide them into chloroplasts or mitochondria are known as transit sequences. A number of sequences act as retention signals, retaining proteins in compartments against a bulk flow of proteins that move out of that compartment without the aid of a particular sequence. Some examples of targeting sequences are listed in *Table 1.1*.

1

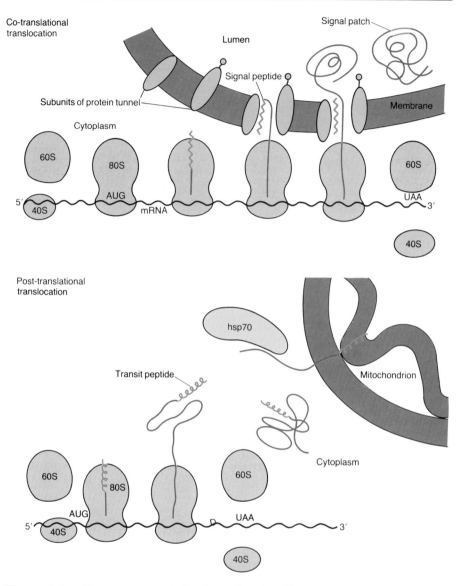

Figure 1.1. Co- or post-translational translocation. Signal peptides may contact the protein translocation machinery in the membrane co-translationally. A signal patch cannot form until the protein folds in the lumen. Transit peptides can bring about translocation post-translationally, but heat-shock proteins are required to unfold the precursor.

Targeting patches are made up of different parts of the structure brought together by the secondary and tertiary structure of the proteins (*Figure 1.1*). For accessibility, they lie on the surface of the protein. Signal patches are implicated in the targeting of proteins that have folded up in the ER lumen. As they are difficult to identify, few examples are known. Cathepsin D, for example,

Table 1.1. Examples of targeting sequences

Translocation into ER lumen

```
+ +    o        o  o   o ^+
MKWVTFLLLLFISGSAFS^RGVF
```

ER stop-transfer sequence

```
 +oo   o              +     +
-KSSIASFFFIIGLIIGLFLVLRVGIH-
```

ER retention sequence (soluble protein)

```
 +--
-KDEL
```

ER retention sequence (membrane protein)

```
 ++o   --++
-RRSFIDEKKMP
```

Translocation into bacterial periplasm

```
+ +     +o         o+o       -- +
MKANAKTIIAGMIALAISHTAMA^DDIK
```

Nuclear targeting

```
 ++  +--
-PKKARED-
```

Chloroplast targeting

```
+         oo  oo        +oo      o++o  o   o  o    + +
MAPAVMASSATTVAPFQGLKSTAGLPVSRRSGSLGSVSNGGRIRC^M-
```

Thylakoid targeting

```
 +  o +-           o
-IKASLKDVGVVVAATAAAGILAGNAMA-
```

Mitochondrial targeting

```
+  o     +       ++o ooo      +
MLSALARPVGAALRRSFSTSAQNN^AKVAVL-
```

IMS targeting

```
 +  o  +   o         o  oo        -
-KLTQKLVTAGVAAAGITASTLLYAD-
```

Peroxisomal targeting

```
 o+
-GSKL
```

Endocytosis

```
 +    + +   o       +oo ---
-KNWRLKNINSINFDNPVYQKTTEDEVII
```

Yeast vacuole

```
 +
-QRPL-
```

Sequences are written in the single letter code. Here, and throughout the text, hydrophobic residues are in bold type, charged residues are indicated + or −, and amino acids with hydroxyl groups on side chains (Thr or Ser residues) are indicated o. ^ marks the position of endopeptidase cleavage.

possesses a signal patch comprising a lysine residue at position 203 in the sequence brought into proximity of sequences between residues 265 and 319 by the folding of the protein, to form a recognition site for N-acetylglucosamine phosphotransferase, an initial step in targeting to the lysosome (Chapter 5, Section 5.2).

3. Cytoplasmic synthesis of proteins

Although a small number of proteins are synthesized in mitochondria (and chloroplasts in plants) the bulk of the cell's proteins are synthesized in the cytoplasm on mRNA that originated in the nucleus. The mechanisms of protein synthesis are complex. Briefly, proteins are synthesized from the amino-terminal end to the carboxy-terminus from ribosomes which move along the mRNA from the 5'-end to the 3'-end. In eukaryotes, mRNA translation starts some way from the 5'-end cap structure at a triplet codon sequence AUG which sets the reading frame, and continues until a stop codon (e.g. UAA) is reached. Initiation requires that Met-tRNA binds to the 40S subunit of the ribosome together with initiation factors and GTP. This initiation complex then binds the mRNA at the cap structure, and scans along until the first AUG is reached (*Figure 1.1*). The 60S subunit binds, and elongation of the polypeptide chain ensues in an ordered manner according to the triplet codon sequences in mRNA. Several ribosomes can simultaneously be engaged in polypeptide synthesis from different positions on one mRNA. Four high energy phosphate groups are split to make each peptide bond, half for charging tRNA with amino acids, half for reactions occurring on the ribosomes.

The nascent polypeptide elongates through a tunnel in the ribosome so that amino-terminal signal sequences emerge on the surface of the ribosome when about 70 amino acids have been polymerized. Thus, contact of the signal sequences with cytoplasmic receptors, or receptors in the membrane, can be made when the polypeptide is over 70 residues in length, and if the ribosome is attached to the membrane, vectorial transfer across the membrane occurs co-translationally while the polypeptide is being elongated (*Figure 1.1*). There is no mechanistic coupling of translation and translocation, however (2). Attachment of the nascent chain to the ribosome is required to prevent the protein precursor from folding up or aggregating (3). On the other hand, signal patches cannot form until the bulk of the protein has been synthesized and folded.

Some proteins move to their destinations after translation (*Figure 1.1*). To translocate across membranes, these precursors are held in an unfolded state by binding to chaperones (see Section 8).

4. The pathways that proteins take through cells

4.1 Pathways in eukaryotes

The main pathways that proteins take through a mammalian cell are depicted in *Figure 1.2*. One pathway originates by movement from the cytoplasm to the

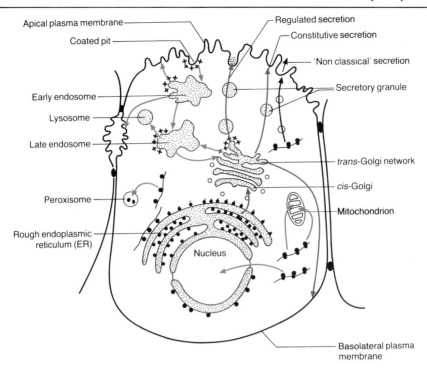

Apical plasma membrane
Coated pit
Early endosome
Lysosome
Late endosome
Peroxisome
Rough endoplasmic
reticulum (ER)
Nucleus
Regulated secretion
Constitutive secretion
'Non classical' secretion
Secretory granule
trans-Golgi network
cis-Golgi
Mitochondrion
Basolateral plasma
membrane

Figure 1.2. Pathways that proteins take through the mammalian cell.

lumen of the ER (see Chapter 3). Some proteins translocate right across the ER
membrane, while only parts of others translocate, leaving the protein integrated
into the membrane (see Chapter 4). Proteins may be either retained in the ER
or moved to the *cis*-Golgi stack, where they may be recycled back to the ER,
or moved on through successive Golgi stacks to the *trans*-Golgi (see Chapter 5).

The *trans*-Golgi is the main sorting station. Its intense vesicular activity deforms
the terminal cisternae into a tubular structure with many buds, known as the
trans-Golgi network. Here, further sorting occurs either to lysosomes via the
endosomal network, or through the constitutive or regulated secretory pathway,
to the cell surface (see Chapter 5). Proteins also move in to the cell from the
plasma membrane, where they are engulfed by endocytosis, travelling through
the endosomal pathway to lysosomes, or back to the same, or a different part,
of the cell's surface. To travel beyond the ER, or from the cell surface, proteins
are transported in vesicles which bud off from the donor membrane, then fuse
with the acceptor membrane (see Chapter 5). The compartments stippled in
Figure 1.2 are topologically equivalent to each other and to the outside of the
cell. Once a protein has crossed the ER membrane, it never has to cross another
membrane even if it is secreted from the cell.

The second major set of pathways of newly synthesized proteins is from the
cytoplasm directly into mitochondria, peroxisomes, chloroplasts, or the nucleus
(see Chapter 6). After entry into mitochondria and chloroplasts, proteins are

sorted further into the various subcompartments of these organelles. From the cytoplasm, proteins may become attached directly to the cytoplasmic surface of membranes, or be secreted by an alternative pathway. Major pathways in eukaryotic cells are summarized diagrammatically in *Figure 1.3*.

4.2 Pathways in prokaryotes

Gram-negative bacteria have an outer and an inner membrane. The outer membrane prevents entry of toxic compounds and allows diffusion of nutrients and ions. The inner cytoplasmic membrane is the major permeability barrier and is the principle location of enzymatic systems including electron transport, oxidative phosphorylation, and lipid biosynthesis. The aqueous environment between the two membranes is known as the periplasm, and contains proteases and phosphatases as well as amino acid and sugar binding proteins that function

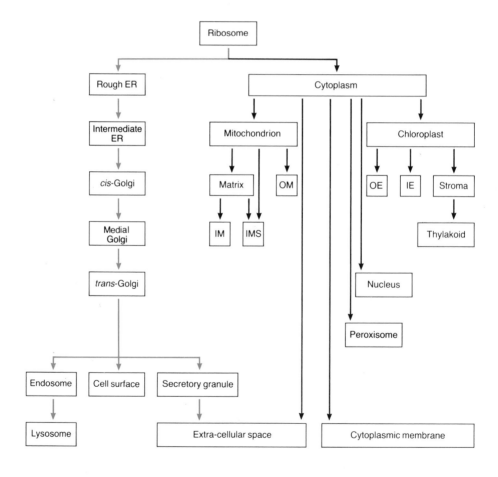

Figure 1.3.   Protein pathways subdivide into two main classes; those that originate from bound ribosomes in the rough ER, and those that originate in the cytoplasm.

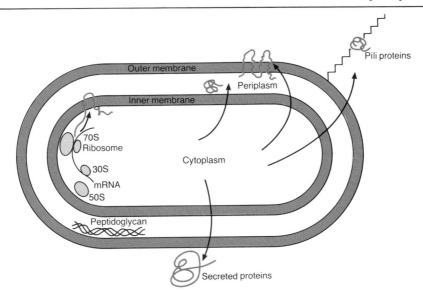

Figure 1.4.   Protein targeting in bacteria.

in active transport. Of the proteins synthesized in the bacterial cytoplasm 25 per cent are targeted either to the inner membrane, the periplasm or the outer membrane (*Figure 1.4*). Several toxins, and amylase from *Bacillus subtilis*, are secreted from the cell into the medium; pili proteins assemble on the outer surface. Most proteins targeted from the cytoplasm are initially synthesized with amino-terminal signal sequences which are recognized by a translocation machinery in the inner membrane (see Chapter 2).

5. Ways of identifying targeting sequences

Targeting sequences are discovered by the use of genetic engineering to create a chimeric protein in which attachment of a suspected targeting sequence to a carrier protein re-routes that protein to a different compartment. Localization of the altered protein is tested after transfecting cells with engineered DNA, expressing the chimeric protein *in vitro* together with the subcellular organelle, or injecting the transcribed mRNA into oocytes. For example, the cDNA encoding an amino-terminal signal sequence of a secretory protein was cut out using restriction enzymes from the full-length DNA and ligated to the 5'-end of the coding DNA of α-globin, appropriately cut and trimmed so that the reading frame was maintained. Again, using restriction enzymes, the fused DNA was then attached on to the 3'-side of a promoter which is recognized by SP6 RNA-polymerase (*Figure 1.5*). When the mRNA, that is produced by RNA polymerase, is translated in a cell-free system together with membrane fragments derived from the ER the translated protein was translocated across the ER membrane (4).

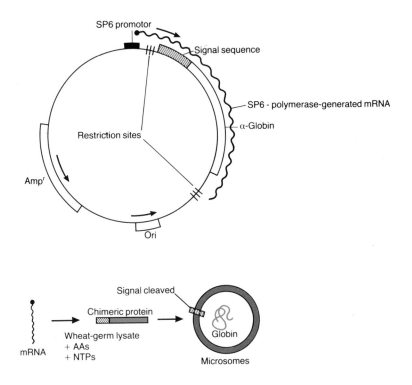

Figure 1.5. Generation of a fused protein. DNA encoding a signal sequence ligated to DNA encoding for a normally cytoplasmic protein is transcribed and translated to yield a precursor protein which is translocated into microsomal vesicles. The signal sequence is cleaved off upon translocation.

6. Evolutionary relationships in targeting

As can be seen in *Table 1.1* the signal sequences that target proteins to the ER membrane, the bacterial inner membrane, the thylakoid membrane, and through the inner membrane of mitochondria into the inter-membrane space (IMS) all have long stretches of hydrophobic residues. Often, a eukaryotic protein expressed in bacteria can be translocated across the inner membrane, and bacterial signal sequences work in eukaryotic cell-free systems (4). The ER is thought to have arisen by invagination of the plasma membrane, so may at one time have been equivalent to the bacterial cytoplasmic membrane.

Mitochondria and chloroplasts, in contrast, are likely to have developed from an engulfed bacteria with whom the original organism lived in symbiosis. Transport of proteins from the matrix to the IMS would thus be topologically equivalent to moving proteins from the bacterial cytoplasm to the periplasm.

7. The role of targeting sequences in protein folding

Gottfried Schatz showed that a chimeric protein containing the amino-terminal 22 residues of a transit sequence from yeast cytochrome subunit IV fused to the enzyme dihydrofolate reductase (DHFR) could not be transported into mitochondria in the presence of methotrexate, a powerful inhibitor of DHFR, which prevented the enzyme unfolding at the surface of the mitochondrion (5). The experiment showed that precursors need to be unfolded to cross membranes. One of the roles of the signal sequence of precursor proteins exported from *E. coli* is to slow refolding compared to that of mature proteins (6), giving the precursor protein time to enter the export pathway while proteins that fold quickly remain in the cytoplasm.

8. Role of chaperone proteins

In the environment that a protein encounters inside the cell, there are many possibilities for incorrect interactions with other macromolecules before it folds into its mature form. Chaperones (*Table 1.2*) are additional proteins that help assembly and folding, but are not part of the final structure. Chaperones form stable complexes with precursor proteins to prevent them folding into mature structures prematurely, and help to maintain them in an open conformation that will go across membranes.

Some chaperones are members of the heat-shock family, that is proteins whose synthesis is increased when the organism is subjected to raised temperatures. They provide hydrophobic surfaces that stabilize proteins which would otherwise denature. One set of chaperones is homologous to an *E. coli* protein known as

Table 1.2 Chaperones

Protein (gene)	Regulation	Source	Mol. mass	Function
GroEL	Heat shock	*E. coli*	57 259	β-lactamase secretion Assembly of phage capsid
GroES	Heat shock	*E. coli*	10 368	β-lactamase secretion
Rubisco ss BP	Heat shock	Chloroplast	61 000	Assembly of Rubisco
Hsp60	Heat shock	Mitochondria	60 000	Assembly of matrix proteins
Hsc70	Cell growth	Cytoplasm	70 000	Uncoating clathrin
Hsp70 (ssa1/2) (ssa3/4)	Constitutive Heat shock	Cytoplasm and nucleus	70 000	Translocation into ER mitochondrion
BiP (kar2)	Glucose lack	ER lumen	78 000	Oligomerization in ER
DnaK	Heat shock	*E. coli*	69 000	Initiates phage DNA synthesis
SSC1	Heat shock	Mitochondria	70 000	Redirecting proteins to IMS
Hsp90	Heat shock	Cytoplasm	90 000	Assembly of p60^{src} plasma membrane

GroEL, while another set is related to a eukaryotic heat-shock protein of 70 kDa mol. mass (hsp70). Eukaryotic organisms generally have several genes encoding proteins related to hsp70.

Chaperones bind ATP, and can be isolated by affinity chromatography on ATP-agarose. They are weak ATPases, and it is thought that unfolded proteins first bind to chaperones and are then released in close proximity to the membrane to which they are targeted with concomitant hydrolysis of ATP. In yeast strains in which hsp70 genes can be repressed, precursors for both the ER lumen and mitochondria accumulate in the cytoplasm (7), and *in vitro* hsp70 has been found to stimulate protein import into yeast, or pancreatic, ER microsomes and mitochondria (8) (*Figure 1.6*).

GroEL is an oligomer of 14 subunits which form two stacked rings of seven subunits each, and associates with GroES which forms one 7-membered ring. It is required for protein secretion (9) and for assembly of the multimeric cyanobacterial ribulose – 1,5 – bis-phosphate carboxylase-oxygenase (Rubisco) in *E. coli* (10). A protein 46 per cent homologous to GroEL is required for assembly of Rubisco in chloroplasts (11) and another member of the family, hsp60, for assembly of matrix proteins in mitochondria (12).

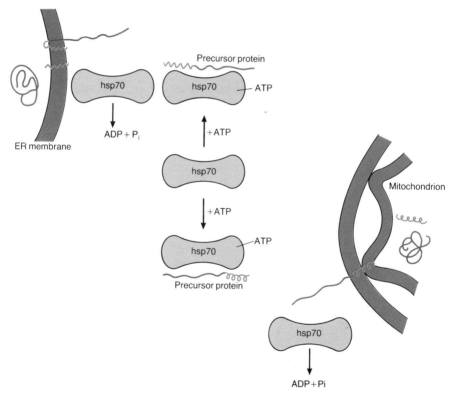

Figure 1.6. Role of heat-shock protein 70 in protein translocation. Together with ATP, the unfolded precursor proteins bind to hsp70, and are released close to the appropriate membrane with concomitant ATP hydrolysis.

9. Further reading

Pelham,H.R.B. (1989) Speculations on the functions of the major heat-shock and glucose-regulated proteins. *Cell*, **46**, 959 – 61.

Meyer,D.I. (1988) Preprotein conformation: the year's major theme in translocation studies. *Trends Biochem. Sci.*, **13**, 471 – 4.

Rothman,J.E. (1989) Polypeptide chain binding proteins: catalysts of protein folding and related processes in cells. *Cell*, **59**, 591 – 601.

Schatz,G. (1986) A common mechanism for different membrane systems (News). *Nature*, **321**, 108 – 9.

10. References

1. Blobel,G. (1980) *Proc. Natl. Acad. Sci. USA*, **77**, 1496 – 1500.
2. Zimmerman,R. and Mollay,C. (1986) *J. Biol. Chem.*, **261**, 12889 – 95.
3. Mueckler,M. and Lodish,H. (1986) *Nature*, **322**, 549 – 52.
4. Lingappa,V.R., Chaidez,J., Yost,C.S. and Hedgpath,J. (1984) *Proc. Natl. Acad. Sci. USA*, **81**, 456 – 60.
5. Eilers,M. and Schatz,G. (1986) *Nature*, **322**, 228 – 32.
6. Park,S., Liu,G., Topping,T.B., Cover,W.H. and Randall,L.L. (1988) *Science*, **239**, 1033 – 5.
7. Deshaies,R.J., Koch,B.D., Werner-Washburne,M., Craig,E.A. and Schekman,R. (1988) *Nature*, **332**, 800 – 5.
8. Chirico,W.J., Waters,M.G. and Blobel,G. (1988) *Nature*, **332**, 805 – 10.
9. Kusukawa,N., Yura,T., Ueguchi,C., Akiyama,Y. and Ito,K. (1989) *EMBO J.*, **8**, 3517 – 21.
10. Goloubinoff,P., Gatenby,A.A. and Lorimer,G. (1989) *Nature*, **337**, 44 – 7.
11. Hemmingsen,S.M., Woolford,C., van der Vies,S.M., Tilly,K. Dennis,D.T., Georgopoulos,C.P., Hendrix,R.W. and Ellis,R.J. (1988) *Nature*, **333**, 330 – 4.
12. Ostermann,J., Neupert,W., Hartl,F.-U. and Horwich,A.L. (1989) *Nature*, **341**, 125 – 30.

2

Bacterial protein translocation

1. Introduction

Proteins are translated within the cytoplasm of Gram-negative bacteria. Some of these proteins are then moved via translocation machinery in the inner membrane to be integrated into the inner membrane, transported across to the periplasm, integrated into the outer membrane, or exported from the cell.

Most translocated proteins are initially synthesized as precursors with amino-terminal signal peptides which act to transfer the precursor protein across the inner membrane. Cleavage by a leader peptidase occurs at the periplasmic face of the inner membrane, then the mature protein moves to its final destination. Translocation is energy dependent, utilizing ATP and the proton-motive force that exists across the inner membrane. The electrical potential and the chemical potential of protons function interchangeably in export, so the precursor proteins are not simply electrophoresed across the membrane.

2. Experimental approaches for determining protein localization

The localization of an endogenous bacterial protein or one expressed from a bacteriophage or plasmid into the inner membrane or periplasm *in vivo* can be assessed by its susceptibility to proteases after removal of the outer membrane. After lysis of the cells, the inner and outer membranes can be separated from each other by centrifugation on sucrose gradients, and analysed for proteins of interest. Care must be taken with interpretation as occasionally newly synthesized proteins, especially chimeric proteins that have aggregated, can behave as dense particles. Another method that can be used for antigenic proteins is immuno-electron microscopy of the proteins labelled with colloidal gold.

In vitro systems have been devised to study translocation across the inner membrane. A DNA-primed *in vitro* assay has been used; translation of bacterial

13

mRNA is generally co-transcriptional, in contrast to eukaryotes where translation is a separate event following transcription. The active extract contains the enzymes and co-factors required for coupled transcription/translation reactions, but, in common with other *in vitro* systems, needs to be supplemented with amino acids and energy sources (see Appendix A3). Alternatively mRNA in a purified form, on free or membrane-bound polysomes, is incorporated into the bacterial cell-free translation assay.

Bacterial inner membranes are extracted in the form of inverted vesicles (see Appendix B3), where the active component for translocation is on the outer surface and proteins pass into the interior. The vesicles are added to the translation extract. As radioactive amino acids are added, the shift in molecular mass, as a result of cleavage by signal peptidase when the translated protein reaches the interior of the added vesicles, can be visualized by autoradiography of the products after separation by SDS-polyacrylamide gel electrophoresis (SDS-PAGE). The protein is also tested for resistance to added proteases since after translocation it becomes inaccessible.

Both co-translational and post-translational translocation of newly synthesized periplasmic proteins have been observed both *in vitro* and *in vivo*. The ATP dependence observed for post-translational translocation *in vitro* is due, at least in part, to the requirement of secA and other chaperones for ATP (see Section 5.1).

3. The structure of prokaryotic signal peptides

As well as its influence on folding of the precursor protein (see Chapter 1), other functions of the signal sequence include recognition by the cytoplasmic and non-cytoplasmic factors involved in protein translocation and interaction with the membrane. Prokaryotic signal peptides, like eukaryotic ones, have three distinct regions, a positively charged amino-terminus (n-region), a hydrophobic core (h-region), and a hydrophilic carboxy-terminus (c-region) (1) (see *Figure 2.1*).

The n-regions may vary in length but have an overall positive charge which may be responsible for binding to the negatively charged phospholipid head groups on the cytoplasmic surface of the membrane. The h-region is rich in hydrophobic residues (Phe, Leu, Val, Ile, Met, Trp), especially those with the potential for forming α-helices, but also with some helix-disrupting amino acids proline and glycine. The length (minimum of 7–8 residues) and overall hydrophobicity of the h-region is critical in its function in translocating the attached polypeptide across the membrane. Natural, or site-specific mutations, that insert a charged residue in certain locations within the hydrophobic region, prevent translocation.

A major, and largely unsolved, question is whether precursor proteins translocate directly through the phospholipid bilayer, or through a protein tunnel. If the former mechanism holds, then the hydrophobic core of the signal sequence is likely to interact with the hydrophobic interior of the bilayer, and may also

Figure 2.1.   The sequence of a signal peptide. The signal sequence of bacteriophage M13 coat protein, indicating the three distinct regions and the ' – 1, – 3 rule' for the amino acids upstream of the cleavage site. n, amino-terminal; h, hydrophobic core; c, carboxy-terminal.

align the signal sequence within the membrane so that signal peptidase cleaves at the correct site as it emerges into the periplasm.

The amino acid sequence of the c-region is generally six residues in length and demonstrates a high degree of stringency at the cleavage site compared with the rest of the signal peptide, conforming to the ' – 1, – 3 rule'. The sequence of the cleavage site is A – X – B (where B is Ala, Gly or Ser; A is any of B or Leu, Val or Ile) (2). The position – 1 which contributes the carboxyl group to the peptide bond that is cleaved is thus restricted to small uncharged residues (Ala, Gly or Ser) in prokaryotes, although cysteine, threonine, glutamine, proline, and leucine are sometimes found in the – 1 position in eukaryotic signal peptides. Larger residues or charged residues are found at the – 2 position. Thus signal peptidase recognizes or prefers certain amino acids for their steric configuration which determines the position for proteolytic cleavage. Mutation of these residues to bulky or charged residues prevents cleavage, but does not effect translocation. Downstream of the cleavage site, at + 2 within the mature sequence, a negative charge in the form of aspartate and glutamate is preferred.

The amino acid sequence dictates the final structure of the peptide as it approaches and is inserted into the membrane. 2D-NMR, CD, and infra-red measurements have shown that wild-type prokaryotic signal sequences adopt β-structures at low lateral pressures, in the presence of the type of phospholipids likely to be encountered in the membrane, and adopt α-helix under high lateral pressures. The results suggested that the peptide may be in β-structure when interacting electrostatically on the aqueous surface of membranes, and in α-helix when inserted (3) (*Figure 2.2*).

Several models for bacterial protein transport have proposed that a loop or turn structure forms transiently during leader peptide insertion and cleavage. In particular, the eight or so hydrophobic residues of the hydrophobic core may straddle the membrane and the c-region may adopt a β-turn in the vicinity of the cleavage site on the periplasmic side of the membrane.

4. The membrane trigger hypothesis and M13 coat protein

Wickner put forward the membrane trigger hypothesis in 1979 as a model for post-translational membrane insertion and translocation of proteins through a lipid phase rather than a proteinaceous pore (4). He proposed that:
(1) signal peptides influence the folding of precursor proteins, so the apolar

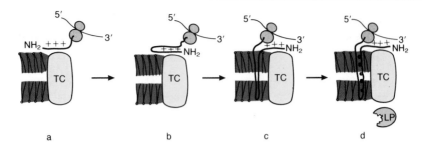

Figure 2.2.   Conformational changes of signal sequences. Changes occurring during translocation; (a) the interaction of the signal peptide with the cytoplasmic surface of the inner membrane; (b) β-sheet formation by signal peptide; (c) insertion of the signal region across the membrane; (d) α-helix formation in the signal peptide. LP, leader peptidase; TC, translocator complex.

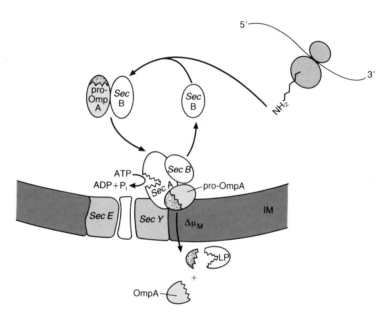

Figure 2.3.   A model for membrane translocation. Translocation of pro-OmpA across the inner bacterial membrane (IM). The location and role of the translocation machinery and membrane electrochemical potential ($\Delta\mu_M$), before cleavage of the leader peptide by leader peptidase (LP) as described in Section 5.2 and *Table 2.1*. By permission of Dr W.Wickner, UC Los Angeles.

residues normally buried in the aqueous environment of the cytoplasm are exposed and available to partition into the hydrophobic bilayer of the membrane;

(2) after insertion, proteins regain their water-soluble conformation and may be extruded into the medium, and the signal peptide is cleaved either co- or post-translocationally;

(3) secretory or membrane proteins can be translocated across or inserted into a membrane post-translationally, without the translocation machinery but utilizing energy in the form of membrane electrochemical potential (*Figure 2.3*).

One protein, the M13 coat protein, follows this model. It becomes integrated into the membrane during the infection stage of the M13 bacteriophage life cycle (5). It is a lipid soluble, 50 residue peptide which is synthesized as a precursor with a 23 residue signal peptide. Information for insertion is present not only within the amino-terminal signal sequence, but also the carboxy-terminal of the mature protein.

In vitro studies show that the M13 procoat protein is processed to its mature form post-translationally (6). Leader peptidase, membrane electrochemical potential and phospholipid are the only components required to mediate binding, processing and insertion of the M13 bacteriophage protein. In contrast, translocation of larger proteins is 'sec dependent', that is dependent on proteins of genes for which isolation of mutants indicates their involvement in protein secretion.

5. Components of the export machinery

Genetic techniques have been applied to identify genes that encode for components of the translocation apparatus. Mutations of one of these, *secY*, which encodes for a protein integrated into the cytoplasmic membrane (7) (*Table 2.1*), was found to reverse the blockage of protein secretion caused by defects in the signal sequences of exported proteins, suggesting that *secY* interacts directly with signal sequences. Exported proteins accumulate at non-permissive temperatures in temperature-sensitive mutants of both *secY* (8) and *secA* (9), a cytoplasmic protein. Knowledge of their individual functions and relationships to one another is not complete but the characterization thus far is presented in *Table 2.1*.

5.1 Prokaryotic chaperones

The role of chaperones in retaining precursor proteins in open conformations prior to translocation was explained in Chapter 1, Section 8. As well as GroEL(S), other proteins with detected chaperone roles include *secA*, *secB*, and 'trigger factor' (10) (see *Table 2.1*).

The precursors for maltose binding protein (MBP), outer membrane protein A (pro-OmpA), and outer membrane porin (pre-PhoE) have been used as model proteins for *in vitro* studies of prokaryotic protein translocation. *SecB*, trigger factor, and GroEL stabilize pro-OmpA for membrane transport and assembly. *SecB* and GroEL seem to stabilize pre-PhoE, but *secB* alone seems to recognize the mature domain of MBP precursor to retard folding and maintain an open conformation (11).

Table 2.1 Components of the bacterial translocation pathway

Component	Mol. mass	Subunit	Location	ATPase activity	Functional Role	Notes
secA	92 kDa	Monomer	Peripheral membrane	Yes	Couples ATP hydrolysis to precursor protein export	Requires secY for ATPase activity
secY	34–42 kDa	Monomer	Integral cytoplasmic membrane	No	Possible signal-sequence receptor; promotes secA activity	Highly hydrophobic molecule, Mr dependent on SDS-PAGE gel system used
secC	ribosomal S15	?	Ribosomal	?	Required for the optimal synthesis of lipoproteins	
secD	40 kDa peripheral domain with up to six trans-membrane domains		Integral cytoplasmic membrane		Required for lipoprotein export in concert with secA, but not secB	
secE	13.6 kDa		Integral cytoplasmic membrane	?		Interacts with secY in translocation
secF	similar size as secD		Integral cytoplasmic membrane	?	Awaits elucidation	

Table 2.1 continued

Component	Mol. mass	Subunit	Location	ATPase activity	Functional Role	Notes
Chaperones						
secB	64 kDa	Tetramer 4 × 16 kDa	Cytosolic	No		Possibly a prokaryotic equivalent of eukaryotic SRP
Trigger factor	63 kDa	Monomeric	Cytosolic	No	Form complexes with export protein precursors to maintain an open and translocation competent conformation	Forms a 1:1 complex with pro-OmpA for *in vitro* membrane assembly: binds 70S ribosomes
GroEL	20S	14 × 65 kDa	Cytosolic	Yes		Involved in phage and oligomeric protein assembly. Belongs to 'heat shock' family of proteins

5.2 A model for protein translocation in Gram-negative bacteria

The secretory machinery used in the prokaryotic system appears to be as intricate as that of eukaryotes. *Figure 2.3* gives a diagrammatical representation of bacterial translocation across the inner membrane in which pro-OmpA is the model protein.

Trigger factor is a 63 kDa chaperone which stabilizes pro-OmpA in an open conformation and has been purified on an affinity column of immobilized pro-OmpA. Another chaperone, *secB*, suggested to be the prokaryotic equivalent of SRP, binds to the signal sequence of the nascent polypeptide and forms a 1:1 complex maintaining the protein in a stable and open conformation for translocation. Once in contact with the membrane, pro-OmpA is then recognized by the peripheral membrane ATPase, *secA*, which binds to both the signal sequence and the mature part of pro-OmpA and couples ATP hydrolysis to release and subsequent translocation of the precursor (12). Release and ATPase activity is stimulated at the membrane by *secY* and the acidic phospholipid components. Pro-OmpA can bind the inner membrane with or without *secA*, but requires *secA* for translocation.

Thought to be the prokaryotic signal sequence receptor, *secY* is an integral membrane protein which potentiates the action of *secA*, and may form part of a proteinaceous pore through which the polypeptide is translocated. It is homologous to a component of yeast ER membrane, *sec61* (see Chapter 3, Section 5). SecY co-purifies with a further membrane component, *secE*. The action of the cytosolic factors, including the chaperones *secB*, *secA*, and its interaction with *secY* occur before membrane insertion of the precursor protein, which is not really understood. Other recently identified *sec* genes, *secC*, *secD*, *secE*, and *secF* are thought to fulfil other fundamental roles in prokaryotic protein export. SecC is thought to be required for optimal lipoprotein synthesis, and *secD* for lipoprotein export in concert with *secA* but not *secB*.

5.3 Comparison of the protein export in Gram-positive and Gram-negative bacteria

A basic structural difference in the Gram-positive bacterium is that they have a one-membrane cell envelope, whereas the Gram-negative bacterium has two membranes. A 64 kDa ribosome-associated protein has been identified in *Bacillus subtilis* which forms a complex with three other proteins of 60, 41, and 36 kDa (13), thus called the 'S complex'. Its very nature would suggest that it is an analogue of eukaryotic SRP (see Chapter 3), but it characteristically resembles *secA*. *Staphylococcus aureus* has a 60 kDa protein which is part of a complex formed with 71, 46, and 41 kDa proteins. It is antigenically related and has a similar cellular distribution to the 64 kDa protein of *B. subtilis* (14).

5.4 Leader peptidase in E. coli

There are two species of the endopeptidase in *E. coli*, leader peptidase I and II. Both are essential for bacterial cell growth but not for translocation. Repression of leader peptidase I leads to accumulation of translocated but improperly folded proteins in the periplasm.

Leader peptidase I is a 36 kDa integral membrane protein which has been purified from detergent extracts of the membranes of an overproducing strain of *E. coli*. The assembly of newly-synthesized enzyme into bacterial membranes has been extensively studied. An uncleaved internal signal sequence, its periplasmic carboxy-terminal, and a polar cytoplasmic domain are all involved, and the *sec* gene products and membrane electrochemical potentials are essential. The cytoplasmic domain is sometimes referred to as the 'translocator poison sequence', a unique membrane assembly element which can block the action of the signal peptide that either precedes or follows it. There is also an apolar domain known as the 'hydrophobic helper' as it is there to overcome the block of membrane assembly by the 'translocator poison' (15).

Leader peptidase II is an 18 kDa protein located exclusively at the inner membrane and is responsible for processing lipoproteins in *E. coli*. It requires a strongly conserved consensus sequence at the cleavage site which is: Leu – Leu – Ala – Gly – Cys. The glycine – cysteine bond is cleaved and a lipid molecule added to the cysteine (16).

6. Translocation from the inner to the outer membrane

The precise mechanism remains unknown for the subsequent pathway for translocation and assembly of proteins into the outer membrane. Three mechanisms which have been proposed (17) are presented diagrammatically in *Figure 2.4*.

Outer membrane proteins are either: (i) exported via vesicular intermediates in the periplasm which 'bud' from the inner membrane, travel through the periplasm, and fuse with the outer membrane (*Figure 2.4*, mechanism A), (ii) translocated across the inner membrane into the periplasm and then integrated into the outer membrane without the apparent requirement of a second signal sequence (mechanism B), (iii) assembled by crossing both the inner membrane and the periplasm either directly, or at Bayer's patches where the inner and outer membrane are at close proximity (mechanism C).

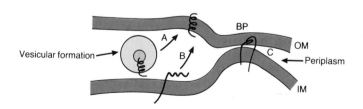

Figure 2.4. Translocation between inner and outer membrane. The three possible mechanisms proposed for the translocation of proteins from the inner membrane (IM) to the outer membrane (OM) in *E. coli*. (A) vesicular formation to enable transport through the periplasm, (B) translocation without the apparent need for a second signal sequence; (C) direct crossing from the inner to the outer membrane via Bayers patches (BP).

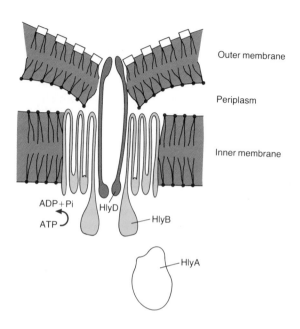

Outer membrane

Periplasm

Inner membrane

Figure 2.5. The novel secretory pathway of haemolysin (HylA). The translocator complex consists of HlyB and HlyD and energy derived from ATP hydrolysis by HlyB.

7. The novel secretion of haemolysin from *E. coli*

E. coli haemolysin (HlyA) is a 107 kDa protein which is secreted independently of an amino-terminal signal sequence and the *secA* gene products (18). HlyA has a 27 amino acid carboxy-terminal sequence containing an area of 16 small amino acids called the 'aspartic acid box'. These residues are important, for they are recognized by a membrane-bound translocator complex composed of HlyB (two polypeptides, 66 and 46 kDa) and HlyD (53 kDa). HlyB is ATP-dependent and couples the energy of ATP hydrolysis to translocation of haemolysin. HlyD is more involved in the act of translocation itself (see *Figure 2.5*).

8. Further Reading

Hardy,S.J.S. and Randall,L.L. (1989) Biochemical investigation of protein export in *Escherichia coli. J. Cell Sci. Suppl.*, **11**, 29–43.
Randall,L.L., Hardy,S.J.S. and Thom,J.R. (1987) Export of protein: A biochemical view. *Ann. Rev. Microbiol.*, **47**, 507–41.

Saier,M.H., Werner,P. and Muller,M. (1989) Insertion of proteins into bacterial membranes: mechanism, characteristics, and comparisons with those of eucaryotes. *Microbiol. Rev.*, **53**, 333–66.

Wickner,W. (1989) Secretion and membrane assembly. (Review). *Trends Biochem. Sci.*, **14**, 280–3.

9. References

1. von Heijne,G. (1988) *Biochim. Biophys. Acta*, **947**, 307–33.
2. von Heijne,G. (1983) *Eur. J. Biochem.*, **133**, 17–21.
3. Briggs,M.S., Cornell,D.G., Dluhy,R.A. and Gierasch,L.M. (1986) *Science*, **233**, 206–8.
4. Wickner,W. (1979) *Ann. Rev. Biochem.*, **48**, 23–45.
5. Ito,K., Date,T. and Wickner, W. (1980) *J. Biol. Chem.*, **255**, 2123–30.
6. Wiech,H., Sagstetter,M., Muller,G. and Zimmermann,R. (1987) *EMBO J.*, **6**, 1011–16.
7. Akiyama,Y. and Ito,K. (1987) *EMBO J.*, **6**, 3465–70.
8. Fandl,J.P. and Tai,P.C. (1987) *Proc. Natl. Acad. Sci. USA*, **84**, 7448–52.
9. Oliver,D.B. and Beckwith,J. (1982) *J. Bacteriol.*, **150**, 686–91.
10. Lecker,S., Lill,R., Ziegelhoffer,T., Georgopoulos,C., Bassford,P.J.Jr., Kumamoto,C.A. and Wickner,W. (1989) *EMBO J.*, **8**, 2703–9.
11. Collier,D.N., Bankiatis,V., Weiss,J.B. and Bassford,P.J. Jr. (1988) *Cell*, **53**, 273–83.
12. Lill,R., Cunningham,K., Brundage,L.A., Ito,K., Oliver,D. and Wickner,W. (1989) *EMBO J.*, **8**, 961–6.
13. Caulfield,M.P., Horiuchi,S., Tai,P.C. and Davis,B.D. (1984) *Proc. Natl. Acad. Sci. USA*, **81**, 7772–6.
14. Adler,L.A. and Arvidson,S. (1984) *FEMS Microbiol. Lett*, **23**, 17–20.
15. von Heijne,G., Wickner,W. and Dalbey,R.E. (1988) *Proc. Natl. Acad. Sci. USA*, **85**, 3363–6.
16. Pollitt,S., Inuoye,S. and Inuoye,M. (1986) *J. Biol. Chem.*, **261**, 1835–7.
17. Baker,K., Mackman,N. and Holland,I.B. (1987) *Prog. Biophys. Molec. Biol.*, **49**, 89–115.
18. Holland,I.B. (1989) *Biochem. Soc. Trans.*, **17**, 828–30.

3

Protein translocation at the rough endoplasmic reticulum

1. Historical background

In the late 1960s Palade and co-workers mapped the export route of proteins from synthesis to extracellular secretion using radioactively-labelled amino acids and time-dependent autoradiography techniques (1). They demonstrated that exported proteins are synthesized on ribosomes attached to the ER membrane

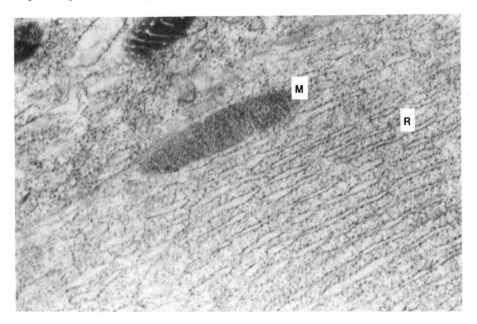

Figure 3.1. Electron micrograph of rough endoplasmic reticulum. Preparation from guinea pig pancreas (× 21 240). Ribosomes (R) can be seen attached to the cytoplasmic surface; (M) mitochondrion.

(*Figure 3.1*) and sequestered to the ER lumen prior to moving to the Golgi apparatus. Redman and his co-workers showed that proteins for export are selectively made on bound ribosomes, while cytoplasmic proteins are synthesized on free ribosomes in the cytoplasm.

The way in which ribosomes translating mRNA coding for secretory proteins become attached to the ER membrane was suggested by the work of Milstein and colleagues. The mRNA for light chain immunoglobulin translated in a cell-free system yielded a larger protein than the mature, secreted form, with an extra piece (called a signal sequence) at the amino-terminal end. The signal sequence was cleaved off when membranes from the ER were present during translation. Blobel and Sabatini then put forward the signal hypothesis which proposed a mechanism for translocation through the ER membrane (2). They proposed that secretory proteins are bound to the ER membrane *via* the signal sequence in the nascent polypeptide chain as it emerges from the ribosome, and that as translation proceeds the nascent polypeptide is extruded through a proteinaceous pore into the ER lumen. The signal sequence is then cleaved by signal peptidase and the completed polypeptide is released from the ribosome which detaches from the ER membrane. When all the secretory protein has reached the lumen, the proteinaceous pore disassembles.

2. The use of *in vitro* systems for studying protein translocation

The development of mRNA-directed protein translation in a cell-lysate, to which membranes can be added, has allowed identification of the molecular entities which regulate the targeting and translocation of proteins to, and across, the ER membrane.

Cell-free translation systems have been prepared from lysates of a number of cell types including wheat-germ, reticulocytes, and yeast (see Appendix A). They contain the ribosomes and additional co-factors required for protein biosynthesis, but need to be supplemented with buffers and energy sources. A source of exogenous amino acid including a radiolabelled amino acid, such as [^{35}S]-methionine, is added. The mRNA for the protein of interest may be extracted from tissues in which it is synthesized. An alternative is to take the cDNA coding for the protein of interest and ligate in to a plasmid downstream of an active promoter. The mRNA may then be generated by *in vitro* transcription using purified RNA polymerase, providing the start codon (AUG) of the protein is the first to be encountered on the 3'-side of the promoter.

Vesiculated portions of the ER (microsomes) are isolated by homogenizing cells such as yeast or pancreas in isotonic buffers, then purified from other membrane components by centrifugation on sucrose density gradients (*Figure 3.2*; see Appendix B). Microsomes are added to a lysate where the mRNA of a secretory protein is being translated, usually during or more rarely, after translation, to observe the processing of the polypeptide that occurs as the vesicles cross to the lumen.

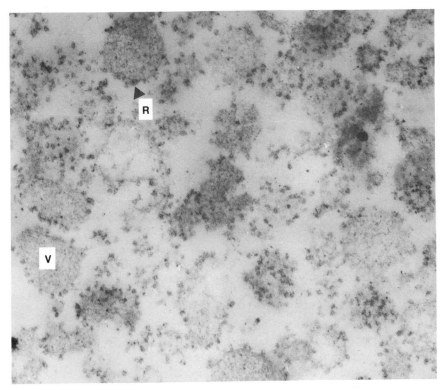

Figure 3.2.   Electron micrograph of microsomes. Rough microsomal vesicles (V) (× 32 480) prepared from pancreas, with ribosomes (R) on the cytoplasmic surface.

The nascent polypeptide chain is targeted and inserted into the lumen of the microsomes where the signal peptide is cleaved off, producing a decrease in the molecular mass of the protein. The products of translation and the effects of translocation may be visualized by SDS-PAGE and autoradiography. *Figure 3.3* shows the $[^{35}S]$-pre-prolactin translated in the absence and presence of microsomes. In the presence of microsomes, cleavage of the signal peptide is demonstrated by a reduced molecular mass compared with the translated precursor, pre-prolactin. Translocation is demonstrated by the addition of proteases as proteins within the microsomes are protected from digestion, unless solubilized with detergent.

3. Structure and function of signal peptides

Secretory and integral membrane proteins (see Chapter 4) are synthesized as precursors, possessing an amino-terminal extension or signal sequence, 15 – 30 amino acids in length. Signal sequences are generally cleaved, although some

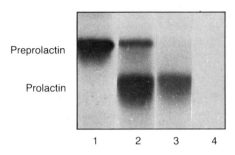

Preprolactin

Prolactin

1   2   3   4

Figure 3.3. *In vitro* processing of pre-prolactin. Autoradiograph of a SDS-polyacrylamide gel visualizing the products of *in vitro* translation and translocation. Lane 1 – pre-prolactin; Lane 2 – pre-prolactin and processing in the presence of rough microsomes to form prolactin; Lane 3 – prolactin protected from protease activity within the microsomal vesicles; Lane 4 – proteolysis of pre-prolactin and prolactin in the presence of proteases and detergent to dissolve the microsomal vesicles.

membrane proteins have signal sequences which are not cleaved after translocation (e.g. cytochrome P-450; see Chapter 4; Section 3.1). Others have internal sequences which act as signal sequences during protein export (e.g. ovalbumin). Signal sequences function to:

(1) maintain the precursor proteins in an unfolded conformation;
(2) interact with signal recognition particle (SRP; see Section 4.1 below);
(3) bring about translocation across the membrane.

The primary amino acid sequence of hundreds of signal sequences are known. Although they share a common function, they display little sequence homology. As in prokaryotes, structures have three characteristics namely:

(1) a hydrophobic core comprising of at least eight uncharged amino acids together;
(2) a flanking polar segment of positively charged residues at the amino-terminus;
(3) small, uncharged residues at the carboxy-terminal cleavage site.

It is the relative positions of the hydrophobic and charged domains, and the secondary structure of the signal peptide which are recognized by the translocation machinery in the ER membrane and by signal peptidase (3). Oligonucleotide-directed mutagenesis has been employed to make changes in the primary sequence of the signal peptide and mature sequence. Introduction of charged residues in the hydrophobic core prevent translocation, and charges in the sequence at the cleavage site can stop recognition by signal peptidase.

Signal sequences may adopt alternative conformations under different environmental conditions, and these changes may facilitate protein export *in vivo*. Signal peptides have been chemically synthesized and their conformational properties in various environments assessed by using a number of biophysical methods such as CD, 2D-NMR, and infra-red spectroscopy.

As a general rule, eukaryotic signal peptides tend to adopt a β-sheet conformation in a highly hydrophobic environment (4). A β-sheet structure within a peptide may allow:

(1) protein – protein interactions between signal peptides and the receptor in the ER membrane;

(2) maximum interaction for minimum length of signal peptide as the signal takes up a transmembrane disposition, with the mature polypeptide looping back across the membrane (5).

In contrast, the prokaryotic signal peptides often form an α-helix which is attributed to the lipid-binding role as well as receptor recognition in the bacterial protein export pathway (see Chapter 2).

4. Proteins involved in eukaryotic translocation

Protein translocation can proceed as a co- or post-translational process. In co-translational translocation, the polypeptide is inserted and translocated through the ER membrane while being synthesized on the ribosome. When performing an *in vitro* assay for translocation the microsomes need to be included as the translation is taking place. When translocation takes place after translation the precursor protein is fully synthesized from the ribosomes before targeting to and insertion into the membrane. Experimentally this is observed when microsomes are added after cycloheximide has been added to halt further translation. Uncoupled from translation, translocation can be shown to require ATP and GTP.

The yeast pheromone, α-mating factor can be processed either co- or post-translationally by yeast microsomes in a yeast cell-free translation system (6). The 'heat-shock' proteins are thought to maintain the precursors in an open or loosely folded conformation prior to translocation, and hydrolyse ATP when releasing the precursors (7). Small proteins, like the bee venom pre-promellitin (70 amino acids), are fully synthesized before emerging from the ribosome therefore translocation must be post-translational and the signal recognition particle cannot impose elongation arrest (see Section 4.1).

4.1 Signal recognition particle and docking protein

Signal recognition particle (SRP) is a rod-shaped 330 kDa ribonucleoprotein complex of six different polypeptides (72, 68, 54, 19, 14, and 9 kDa) and a 7SL RNA (300 nucleotides) isolated from salt extracts of microsomal membranes (8). The 19 kDa subunit is attached directly to the 7SL RNA and mediates binding of the 54 kDa protein. There are two heterodimers, the 72/68 kDa and the 14/9 kDa subunits (9) (*Figure 3.4*). The SRP was originally isolated because of its activity in reconstituting processing activity to salt-extracted microsomes. In cell-free lysates it is found to halt or slow down the elongation of secretory proteins unless microsomes are present. Thus *in vivo*, the SRP is thought to prevent cytoplasmic synthesis of a secretory protein by holding up completion until the ribosome has made contact with the ER membrane.

SRP-mediated translocation only acts when nascent chains are still attached to the ribosome. The 54 kDa subunit possesses a GTP-binding site, and has a

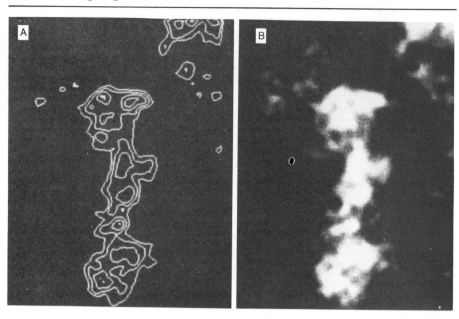

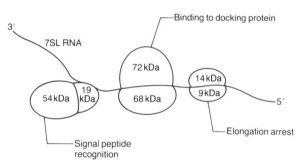

Figure 3.4. **(a)** Morphology of the signal recognition particle (SRP). Contour image of density (molecular mass) (A) and a dark field electron micrograph (B) of SRP. By permission of Dr David Andrew of McMaster University, Canada. **(b)** The orientation of the protein subunits and 7SL RNA within the SRP complex.

methionine-rich carboxy-terminal domain thought to be folded into several amphiphilic helices which bind the signal sequence (10) as it emerges from the ribosomes when the nascent polypeptide chain is around 70–80 amino acids in length. Elongation arrest is imposed by the 14/9 kDa dimer until the 68/72 kDa dimer binds to the docking protein in the ER membrane, at which point translation is resumed (see *Figure 3.5*). The 7SL RNA has a highly conserved structural domain, whose level of abundance across species in eukaryotes and prokaryotes would suggest a common evolutionary origin and function. The functional role of the RNA is to maintain all the other protein subunits of the SRP complex in a stable conformation and form specific base pair interactions with rRNA and mRNA.

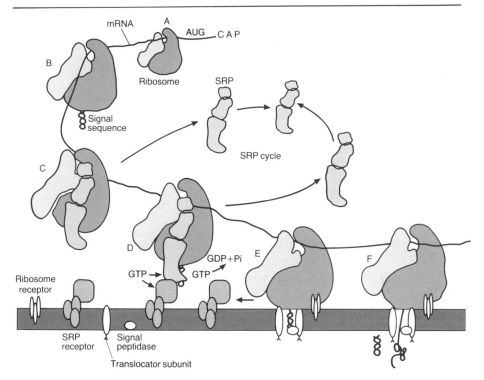

Figure 3.5.   Role of the signal recognition particle (SRP). (A) Protein synthesis from the mRNA by the ribosomes; (B) signal sequence emerging from the ribosomal subunits; (C) signal sequence interaction with SRP, halting elongation; (D) SRP interaction with its receptor, docking protein, at the ER membrane, SRP released from the ribosome and translation resumed; (E) interactions of the ribosome with its receptor on the ER membrane, signal sequence with the signal sequence receptor and translocator proteins forming an aqueous pore for translocation; (F) nascent peptide translocated through the aqueous pore and signal sequence cleaved by signal peptidase.

Docking Protein (DP) or SRP receptor is an integral membrane protein of the rough ER, made up of two subunits, SRalpha (72 kDa) and SRbeta (30 kDa) (11, 12). The 72 kDa subunit has a 60 kDa cytoplasmic domain which can be released from the membrane by protease digestion, and a 12 kDa domain which has an anchorage role, to the SRbeta subunit. SRalpha interacts with SRP to release the translation arrest by displacing SRP from the ribosome (see *Figure 3.5*).

SRalpha had a GTP-binding site in its carboxy-terminal cytoplasmically disposed domain which shares sequence homology with the SRP 54 kDa subunit. GTP binding is required before SRalpha can catalyse the displacement of SRP from the ribosome, a step which has to occur before the signal sequence is released from SRP and the polypeptide is free to insert into the translocation complex (13). GTP hydrolysis may be required for the release of SRP from SRalpha. Docking protein has no affinity for ribosomes, so that additional components of the ER are involved in maintaining a functional ribosome-membrane junction.

4.2 Ribosome binding proteins

Ribophorins I and II are glycoproteins of the rough ER membrane which have been proposed to be involved in ribosome binding. However, since most of their structure is located in the ER lumen, and are consequently protected from protease digestion, their topography would dispute this role (14). Recently, an integral membrane protein of around 180 kDa has been found to have ribosome-binding activity (15). This protein has a large cytoplasmic domain and can be cross-linked chemically to ribosomes.

4.3 Translocator proteins

Photoreactive diaziridine groups have been incorporated into lysine residues at selected positions in a secretory nascent polypeptide chain during biosynthesis from a diaziridine-modified Lys-tRNA derivative (16). Cross-linkage to the 54 kDa subunit of SRP is formed from these derivatives. Both short nascent chains attached to the ribosome and longer chains are cross-linked to a proteinaceous integral membrane translocator glycoprotein (39 kDa). The subunit is cross-linked to lysine residues either within the amino-terminal and/or the mid-portion of the nascent chain, as it goes through into the ER lumen.

Thus, the 39 kDa glycoprotein (α-subunit) and a smaller 22 kDa protein (β-subunit) which attaches to it, may comprise part of the proteinaceous tunnel through which proteins travel to the ER lumen. Proteins exhibiting cross-reactivity with antibodies to the α subunit have been found in several species, which suggest a conserved function.

An additional membrane protein involved in translocation is known to possess a sulphydryl group, since it is sensitive to N-ethylmaleimide (NEM). A 43 kDa integral microsomal membrane protein has also been identified by photo-cross-linkage to a chemically synthesized signal peptide (17). Binding to the 43 kDa protein is saturable *in vitro* with excess synthetic signal peptides, and the binding of signal peptides to the receptor blocks translocation of nascent chains within *in vitro* systems. This 43 kDa protein enables pre-proteins to select the ER membrane rather than other membranes in the cell, and is especially useful for proteins that are translocated post-translationally, and do not require SRP for targeting. The 43 kDa protein has a 50 kDa homologue in yeast microsomes, and is likely to form part of the translocator complex (*Figure 3.6*) which may have more components still to be discovered (see also Section 5).

A common view is that the translocation complex aids the transfer of the hydrophilic protein across the hydrophobic membrane by providing an aqueous proteinaceous pore. Singer *et al.* presented a model in which translocator proteins have a specific structural organization within the ER (18). The proposed roles for these translocator proteins include:
(1) binding to the signal sequence of the nascent polypeptide;
(2) forming a complex which functions as an aqueous transmembrane tunnel.

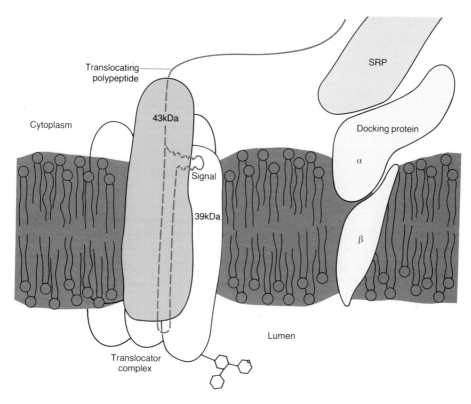

Figure 3.6.   A model for protein translocation. The translocation complex may consist of a 43 kDa signal-binding subunit, a 39 kDa glycoprotein, a 22 kDa protein, and homologues of *sec61, 62,* and *63* gene products. Interaction with the translocator complex occurs after targeting of the nascent polypeptide to the ER membrane via SRP and DP.

So subdomains of the nascent polypeptide pass down the central axis of the transmembrane tunnel until the whole peptide is translocated (*Figure 3.6*).

4.4 Signal peptidase

Signal peptidase is an integral membrane endopeptidase which cleaves the amino-terminal signal sequence from nascent or recently completed protein precursor as it emerges into the lumen of the ER. The purified complex from hen oviduct is made up of two subunits, a 19 kDa protein and a glycoprotein which varies in molecular mass depending on the extent of glycosylation (22–23 kDa) (19). Signal peptidase is isolated from canine pancreatic microsomes as a complex of six non-identical units of sizes 25, 23, 22, 21, 18, and 12 kDa, two of which are glycosylated forms of the same protein subunit (subunits 22 kDa and 23 kDa) (20). Some of these proteins may be involved in other aspects of translocation. The exact mechanism by which signal peptidase cleaves the amino-terminal signal sequence has not been elucidated. The peptidase solubilized with detergent requires phospholipids for activity.

5. ER of yeast

Yeast cells secrete a small number of proteins, of which α-mating factor and invertase have been the most studied. The pathway of protein secretion in yeast appears to resemble the pathways in plants and animals, although less ER, Golgi apparatus, and secretory vesicles are seen by electron-microscopy. These observations are consistent with rapid transit time and low levels of precursors. There is a correlation between the extent of protein secretion and the amount of cell surface growth (budding), so it is assumed that secretory mutants would be lethal. It is possible, however, to isolate temperature-sensitive *sec* mutants that grow normally at the permissive temperature (20–25°C), but fail to secrete at the non-permissive temperature (37°C) (21).

Yeast secretory mutants have been selected because they synthesize, but fail to export, active invertase and acid phosphatase at non-permissive temperatures. Analyses of the cellular pool of precursor secretory proteins are used to evaluate the point in the secretory pathway at which mutants are blocked, causing an accumulation of these secretory enzymes in organelles. At the genetic and protein level these mutants can be used to analyse the function of the gene products in secretion.

Secretory mutants, defective in ER translocation that have been characterized include *sec61, sec62, sec63* (22,23), and *sec11*. The *sec61* gene product (*sec61p*) is a polytopic integral membrane protein with at least five membrane spanning domains. Although its molecular mass, according to its cDNA sequence, is 51 kDa, it runs on electrophoresis with an apparent molecular mass of around 38 kDa, determined by SDS-PAGE, owing to its hydrophobic character. *Sec61p* exhibits amino acid sequence homology with the bacterial *sec* gene product *secY* (see Chapter 2) and may comprise part of the translocator complex.

Sec62 gene product is also an integral membrane protein whose DNA sequence has revealed molecular mass of 32 kDa with no obvious amino-terminal signal sequence but two possible membrane spanning domains. The 30 amino acids of the carboxy-terminal could form an α-helix which may allow *sec62p* to interact with other proteins involved in translocation.

The *sec63* gene product is a third membrane-associated component of the ER translocation machinery in yeast. It is thought the *sec61, sec62,* and *sec63* form a complex together with a 31.5 kDa glycoprotein and a 23 kDa protein, homologues of the mammalian α and β-subunits that cross-link to nascent chains and act in tandem in polypeptide translocation. Two defective or partially functioning proteins will exaggerate the overall deficiency in a phenomenon known as synthetic lethality.

Sec11 gene product has physico-chemical properties which coincide with that of a predicted signal peptidase lesion in yeast (24). It is a basic 18.8 kDa protein with a pI of 9.5, has a potential N-glycosylation site, and is homologous in sequence to the 21 kDa and 18 kDa subunits of canine signal peptidase. The *sec11* mutant yeast strain accumulates unprocessed forms of glycoproteins in the ER lumen.

Genes which are homologous to the 54 kDa and 19 kDa subunits of the SRP and to SRalpha have been sequenced in yeast, but functions of the corresponding proteins have not been identified. An additional protein required for yeast translocation across the yeast ER membrane is hsp70 which facilitates the post-translational import into the ER by preventing the precursor proteins from folding up (see Chapter 1; Section 8).

6. Further reading

Robinson,A. and Austen,B.M. (1987) The role of topogenic sequences in the movement of proteins through membranes. *Biochem. J.*, **246**, 249–61.
Rothblatt,J. and Schekman,R. (1989) A hitchhiker's guide to analysis of the secretory pathway in yeast. *Methods in Cell Biology*, **32**, 3–36.
Rothman,J.E. (1989) GTP and methionine bristles. *Nature*, **340**, 433–4.
Briggs,M.S. and Gierasch,L.M. (1986) Molecular mechanisms of protein secretion: the role of the signal sequence. *Adv. Protein. Chem.*, **38**, 109–80.

7. References

1. Palade,G. (1975) *Science*, **189**, 347–57.
2. Blobel,G. and Sabatini,D.D. (1971) In *Biomembranes*, Ed. L.A. Manson, Plenum Publishing Corp., New York, **2**, 193–5.
3. von Heijne,G. (1985) *Curr. Top. Membr. Transp.*, **24**, 151–79.
4. Fidelio,G.D., Austen,B.M., Chapman,D. and Lucy,J.A. (1987) *Biochem. J.*, **244**, 295–301.
5. Austen,B. (1979) *FEBS Letts*, **103**, 308–12.
6. Adcock,I.M., Westwood,O.M.R. and Austen,B.M. (1989) *Biochem. Soc. Trans.*, **17**, 823–7.
7. Deshaies,R.J., Koch,B.D., Werner-Washburne,M., Craig,E.A. and Schekman,R. (1988) *Nature*, **332**, 800–5.
8. Walter,P. and Blobel,G. (1980) *Proc. Natl. Acad. Sci. USA*, **77**, 7112–16.
9. Andrew,D.W., Walter,P. and Ottensmeyer,F.P. (1985) *Proc. Natl. Acad. Sci. USA*, **82**, 785–9.
10. Kurzchalia,T.V., Weidmann,M., Girshovich,A.S., Bochkareva, E.S., Bielka,H. and Rapoport,T. (1986) *Nature*, **320**, 634–6.
11. Meyer,D.I. and Dobberstein,B. (1980) *J. Cell Biol.*, **87**, 503–8.
12. Andrew,D.W., Lauffer,L., Walter,P. and Lingappa,V.R. (1989) *J. Cell Biol.*, **108**, 797–810.
13. Connolly,T. and Gilmore,R. (1989) *Cell*, **57**, 599–610.
14. Crimaudo,C., Hortsch,M., Gausepohl,H. and Meyer,D.I. (1987) *EMBO J.*, **6**, 75–82.
15. Savitz,A.J. and Meyer,D.I. (1990) *Nature*, **346**, 540–4.
16. Weidmann,M., Kurzchalia,T.V., Hartmann,E. and Rapoport,T. (1987) *Nature*, **328**, 830–3.
17. Robinson,A., Kaderbhai,M.A. and Austen,B.M. (1987) *Biochem. J.*, **242**, 767–77.
18. Singer,S.J., Maher,P.A. and Yaffe,M.P. (1987) *Proc. Natl. Acad. Sci. USA*, **84**, 1015–19.
19. Baker,R.K. and Lively,M.O. (1987) *Biochemistry*, **26**, 8561–7.
20. Evans,E.A., Gilmore,R. and Blobel,G. (1986) *Proc. Natl. Acad. Sci. USA*, **83**, 581–5.
21. Novick,P. and Schekman,R.W. (1979) *Proc. Natl. Acad. Sci. USA*, **76**, 1858–62.
22. Rothblatt,J.A., Deshaies,R.J., Sanders,S.L., Daum,G. and Schekman,R.W. (1989). *J. Cell Biol.*, **109**, 2641–52.
23. Deshaies,R.J. and Schekman,R.W. (1989) *J. Cell Biol.*, **109**, 2653–64.
24. Bohni,P.C., Deshaies,R.J. and Schekman, R.W. (1988) *J. Cell Biol.*, **106**, 1035–42.

4

Assembly of membrane proteins

1. Mode of attachment of proteins to membranes

Proteins become attached to membranes either by consecutive stretches of hydrophobic uncharged amino acid residues, or by covalently attached fatty acids. A common criterion for an integrated membrane protein is that it requires detergent for its release from a membrane, whereas it is not released by high pH (pH 11.5) (1). Proteins linked through ester linkages to fatty acids are released by 1 M hydroxylamine or high pH.

2. Topography of integral membrane proteins

Transmembrane proteins may either span the phospholipid bilayer once, twice, or many times. Topology is established when the protein first assembles into the membrane. For plasma membrane proteins, topology is adopted at the ER membrane, as subsequent movement to other organelles or to the cell surface involves vesicle budding and fusion which does not change topology (see Chapter 5, Section 2). Thus, a portion of membrane protein that remains cytoplasmic at the ER membrane, is still cytoplasmic after the protein has been moved to the plasma membrane. Membrane proteins are also assembled into the membranes of mitochondria, chloroplasts, and other organelles after synthesis in the cytoplasm.

The first three-dimensional model of a membrane protein obtained from electron micrographs of the purple membrane of *Halobacterium halobium*, showed that bacteriorhodopsin contained seven α-helices spanning the membrane (2). A full determination by X-ray crystallography has been performed only for the complex photosynthetic reaction centre of the bacterium *Rhodopseudomonas viridis* (3), and this also shows a number of transmembrane helices.

The transmembrane portions of membrane protein contain about 19 hydrophobic uncharged residues in one stretch and all molecules are arranged the same way round (4). The hydrophobic interior of the membrane bilayer is a thermodynamically unfavourable environment unless all the peptide bonds are involved in hydrogen bonding. Complete hydrogen bonding occurs only in the α-helix, so this is the most stable membrane-spanning structure, although β-sheets are found in the transmembrane region of cytochrome b_5. Multi-spanning membrane proteins often act as ion channels or gates, and in consequence contain a number of amphiphilic helices positioned so that their hydrophobic surfaces face the phospholipids, and their hydrophilic surfaces face inwards to generate an aqueous channel.

2.1 Methods for determination of topography

The positions of transmembrane domains in proteins are often predicted from the distribution of hydrophobicity (5). Exposed regions can be detected by their accessibility to reaction with charged chemical probes such as the fluorescent compound 3-azido-2,7-naphthalene disulphonate or by proteases or antibodies which do not cross the membrane (4). Regions of proteins that contain covalently-linked oligosaccharides are assumed to be lumenal, or on the cell's surface (see Chapter 5, Section 2.1). In bacteria, topology can be determined from the cellular location of enzyme activity of alkaline phosphatase fused to portions of membrane proteins (6).

3. Mechanisms of assembly

3.1 Assembly of single-span proteins

When mRNAs encoding membrane proteins are translated *in vitro* together with microsomal membranes and SRP, proteins are assembled with the same final orientation as found *in vivo*, as summarized in *Table 4.1* (7). The SRP interacts with signal sequences and targets the ribosomes to the membrane as described in Chapter 3. The orientation of single-span proteins depends on the positions in the sequences of the signal sequence and the hydrophobic stretches that when assembled span the membrane.

Membrane proteins are classified into three types, classes 1, 2, and 3. Class 1 includes glycophorin A, the G-protein encoded by vesicular stomatitis virus (VSV-G), the light-density lipoprotein receptor, and the mannose-6-phosphate receptor (see Chapter 5). This class is initially synthesized with a cleavable signal sequence, and possesses an internal hydrophobic domain which ends up spanning the membrane. Class 1 proteins are assembled with their amino-termini on the lumenal side, and their carboxy-termini on the cytoplasmic side. The signal sequence, targeted to the membrane by SRP, inserts into the translocation complex, then, as the nascent chain grows it moves through the complex until the hydrophobic spanning sequence, the stop transfer sequence, is reached. At this point, transfer ceases, synthesis of the cytoplasmic tail is completed, and the polypeptide moves from the translocation complex into the bilayer (*Figure 4.1*).

Table 4.1 SRP-dependent insertion of membrane proteins

Protein	No. of spans	Location of: N-term.	Location of: C-term.	Signal sequence cleaved	SRP-translation dependence arrest
Ca^{2+}-ATPase	7	in	out	−	+	+
Lens MP26	6	in	out	−	+	−
Glycophorin A	1	out	in	+	+	+
VSV-G protein	1	out	in	+	+	+
Cytochrome P-450	1	out	in	−	+	−
Sindbis PE_2	1	out	in	−	+	+
Opsin	7	out	in	−	+	+
Asialoglycoprotein receptor	1	in	out	−	+	+
Glucose transporter	12	in	out	−	+	−
Invariant chain	1	in	out	−	+	+
Transferrin receptor	1	in	out	−	+	+
Influenza A; M_2	1	out	in	−	+	−
Cytochrome b_5	1	in	out	−	−	−

Class 2 and 3 proteins lack cleavable signal sequences but have stop-transfer sequences that double as signal sequences which interact with SRP and enter the translocation complex (8), but with different orientations. Class 2 proteins have a higher concentration of positively charged residues on the amino-terminal side of the stop-transfer sequence than on the carboxy-terminal side (9). Binding of these basic residues to the cytoplasmic surface of the translocation complex while the carboxy-terminal end crosses to the lumen gives rise to the orientation with carboxy-terminus translocated, as shown in *Figure 4.1*. Class 2 proteins include sialyl transferase, sucrase – isomaltase, the asialoglycoprotein receptor, and the transferrin receptor. Class 3 proteins have a number of positively charged residues on the carboxy-terminal side of the stop-transfer sequence, and so end up in the opposite orientation, with their amino-termini in the lumen, and their carboxy-termini in the cytoplasm (*Figure 4.1*). This class includes cytochrome P-450 and epoxide hydratase.

Class 1 and 2 proteins are differentiated by the charged nature of the sequences that flank the hydrophobic domains. Changing these flanking regions can convert a Class 2 protein to one that is cleaved by signal peptidase, and does not remain in the membrane (10).

3.2 Assembly of complex proteins

Proteins such as the Band 3 anion transporter of the red cell membrane, the glucose transporter, and rhodopsin span the membrane many times. Opsin, for example, is a 348-residue membrane protein which together with the visual pigment retinal comprises the light receptor protein rhodopsin in rod cells of the vertebrate retina. The polypeptide crosses the membrane as seven α-helices. The mode of assembly has been investigated by preparing fused proteins containing the amino-terminal 34 residues, which cross the membrane to become

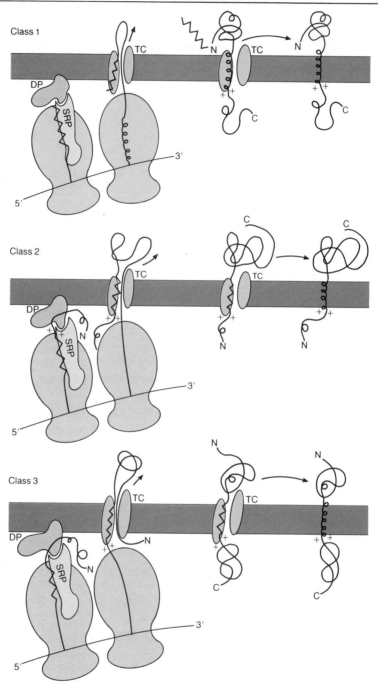

Figure 4.1.   Mechanisms of assembly of single transmembrane proteins. The three different modes of assembly for proteins involve interactions of the signal sequence or stop-transfer sequence with signal recognition particle (SRP), docking protein (DP) and the translocation complex (TC).

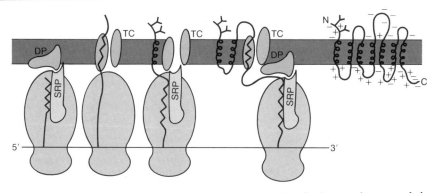

Figure 4.2.   Mechanism of membrane assembly of opsin. Opsin contains several signal sequences, each of which interact with SRP and the translocation complex leading to stepwise assembly.

glycosylated, the carboxy-terminal domain, which remains cytoplasmic, and one selected transmembrane domain. Each transmembrane domain independently interacts with SRP and is inserted in the membrane, presumably into the translocation complex (*Figure 4.2*) (11). The orientation of the amino-terminal segment is determined by the high number of positively charged residues on the carboxy-terminal side of the first stop-transfer sequence (*Figure 4.2*), which fixes the orientation of successive transmembrane loops.

4. Hydrophobic modifications of proteins

Some proteins are targeted to membranes by specific covalent attachment of fatty acids, prenyl groups, or inositol glycolipid (*Figure 4.3*).

4.1 Fatty acid acylation

It was noted that radioactive fatty acids become incorporated into some membrane proteins (12). The fatty acids palmitate, oleate, and stearate become attached to cysteine residues in a thio-ester linkage. Many palmitylated proteins are already integrated into membranes through transmembrane polypeptides (*Figure 4.3*), so it is not clear what the function of the fatty acid is. The VSV-G protein will anchor, for example, when the cysteine acceptor residue for fatty acylation is deleted by mutation Palmitylation of some viral proteins may be required for their fusogenic properties. Palmitylation occurs just as proteins are leaving the ER.

4.2 Prenylation

p21^{ras} is modified by the farnesyl group on residue Cys-186 at the carboxy-terminus. The same modification is found in the yeast a-mating factor (Chapter 5, Section 8.1), and occurs at a consensus sequence Cys – Ala – Ala – X (13).

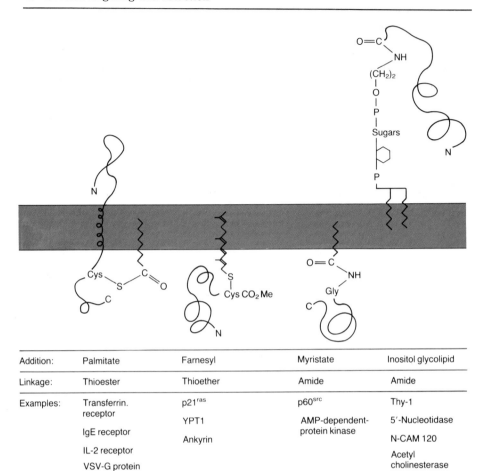

Figure 4.3. Covalent modifications which attach proteins to membranes.

Addition:	Palmitate	Farnesyl	Myristate	Inositol glycolipid
Linkage:	Thioester	Thioether	Amide	Amide
Examples:	Transferrin. receptor	p21^{ras}	p60^{src}	Thy-1
	IgE receptor	YPT1	AMP-dependent- protein kinase	5′-Nucleotidase
	IL-2 receptor	Ankyrin		N-CAM 120
	VSV-G protein			Acetyl cholinesterase

Processing includes cleavage of the three carboxy-terminal amino acids, and carboxymethylation of the carboxy-terminal cysteine residue. In the *ras* oncogene-encoded protein, a small GTP-binding protein, the modification serves to attach the protein to the cytoplasmic surface of cellular membranes. Other cysteine residues in p21^{ras} are also palmitylated.

4.3 Myristoylation

A small group of virally encoded proteins and also p60^{src} oncogene are myristoylated via an amide linkage to an amino-terminal glycine. The modification also serves to attach the protein to a cytoplasmic membrane surface, but less strongly than with palmitylation. Myristoylation occurs by donation from myristoyl-CoA co-translationally in the cytoplasm at the consensus sequence Met – Gly – X – X – X – Ser(Thr) or (Gly).

4.4 Glycosyl-phosphatidylinositol anchors

Some cell-surface proteins are attached by the carboxy-terminal residue to ethanolamine in an amide linkage which is then linked via glycans to phospholipids. These proteins are released into the extra-cellular environment when required by PI-specific phospholipase C (PLC), and it was the lability of these proteins to PLC that first indicated this type of membrane anchor (14). The proteins are initially integrated by a stop-transfer sequence into the membrane in the ER, and then transferred to the phospholipid anchor.

5. Further reading

Jennings,M.L. (1989) Topography of membrane proteins. *Ann. Rev. Biochem.*, **58**, 999 – 1027.

Singer,S.J., Maher,P.A. and Yaffe,M.P. (1987) On the transfer of integral proteins into membranes. *Proc. Natl. Acad. Sci. USA*, **84**, 1960 – 4.

von Heijne,G. (1988) Transcending the impenetrable: how proteins come to terms with membranes? (Review). *Biochem. Biophys. Acta*, **947**, 307 – 33.

Glomset,J.A., Gelb,M.H. and Farnsworth,C.C. (1990) Prenyl proteins in eukaryotic cells: a new type of membrane anchor. *Trends Biochem. Sci.*, **15**, 139 – 42.

Low,M.G. (1989) The glycosyl-phosphatidylinositol anchor of membrane proteins. (Review). *Biochem. Biophys. Acta*, **988**, 427 – 54.

Schmidt,M.F.G. (1989) Fatty acylation of proteins. *Biochem. Biophys. Acta*, **988**, 411 – 26.

6. References

1. Fujiki,Y., Hubbard,A.L., Fowler,S. and Lazarow,P.B. (1982) *J. Cell. Biol.*, **93**, 97 – 102.
2. Henderson,R. and Unwin,P.N.T. (1975) *Nature*, **257**, 28 – 32.
3. Deisenhofer,J., Epp,O., Miki,K., Huber,R. and Michel,H. (1985) *Nature*, **318**, 618 – 24.
4. Bretscher,M.S. (1971) *Nature New Biol.*, **231**, 229 – 32.
5. Kyte,J. and Doolittle,R.F. (1982) *J. Mol. Biol.*, **157**, 105 – 32.
6. Boyd,D., Manoil,C. and Beckwith,J. (1987) *Proc. Natl. Acad. Sci. USA*, **84**, 8525 – 9.
7. Katz,F.N., Rothman,J.E., Lingappa,V.R., Blobel,G. and Lodish,H. (1977) *Proc. Natl. Acad. Sci. USA*, **74**, 3278 – 82.
8. Spiess,M. and Lodish,H.F. (1986) *Cell*, **44**, 177 – 85.
9. von Heijne,G. and Gavel,V. (1988) *Eur. J. Biochem.*, **174**, 671 – 8.
10. Lipp,J. and Dobberstein,B. (1986) *Cell*, **46**, 1103 – 12.
11. Friedlander,M. and Blobel,G. (1985) *Nature*, **318**, 338 – 43.
12. Schmidt,M.F.G. and Schlesinger,M.J. (1979) *Cell*, **17**, 813 – 19.
13. Kamiya,Y., Sakurai,A., Tamura,S. and Takahashi,N. (1978) *Biochem. Biophys. Res. Commun.*, **83**, 1077 – 83.
14. Ikezawa,H., Yamenegi,M., Taguchi,R., Miyashita,T. and Ohyabu,T. (1976) *Biochem. Biophys. Acta*, **450**, 154 – 64.

5

Protein sorting and maturation

1. Regulated and bulk flow

In Chapter 1, the pathways that proteins follow after translocation to the ER lumen were outlined. Transport between subcellular organelles takes place in vesicles. Targeting to lysosomes, or to secretory granules which exocytose under a regulatory stimulus, is mediated by receptors that bind to specific sequences, as is retention of proteins in the ER lumen or in the Golgi. In contrast, secretory and membrane proteins that are moved to the cell surface in a non-regulated constitutive manner move by non-selective bulk flow. Antibody-secreting lymphocytes are an example of cells that constitutively secrete proteins (1). There is no pool of stored immunoglobulin, and the newly synthesized proteins move to the cell surface soon after synthesis. Exocytosis occurs without an external stimulus.

2. Export from the ER

2.1 Protein processing

The rate of constitutive secretion of a small fatty-acid acylated glycopeptide has been found to be between 5 and 20 minutes. Many constitutive proteins move more slowly than this (2), and the rate-determining step is the rate at which proteins leave the ER. Export is determined by the rate that proteins fold and form oligomers in the ER. Misfolded, aggregated, and incorrectly assembled proteins are not allowed to leave the ER and are degraded (3). Prevalent acidic Ca^{2+}-binding proteins present in the ER lumen provide an environment conducive to accurate disulphide pairing and oligomerization, viz.

1. Protein disulphide isomerase (PDI) (EC 5.3.4.1) is a dimeric enzyme ($2 \times$ 57 kDa mol. mass) which with help from oxidized glutathione catalyses the formation of disulphide bonds in or between nascent polypeptide chains

(4). PDI contains two domains which are homologous in sequence to a small bacterial redox protein thioredoxin.

2. Heavy chain **binding protein (BiP)** is a 78 kDa protein related in structure to hsp70 (see Chapter 1, Section 8), but its initial translated form contains a signal sequence. It binds to the hydrophobic domains of single subunits of oligomeric proteins such as IgM or influenza virus haemagglutinin, and protects them from aggregation or interaction with the wrong proteins until the other subunits required are available (3). With concomitant hydrolysis of ATP, BiP then releases subunits so that formation of the oligomer can take place.

2.2 Retention and retrieval of proteins in the ER

ER resident proteins such as PDI and BiP accumulate in the ER by targeting sequence interactions involving their carboxy-terminal sequences, Lys – Asp – Glu – Leu (KDEL) (5). In yeast, the appropriate ER-retention sequence is His – Asp – Glu – Leu (HDEL). Attachment of KDEL to the carboxy-terminus of a normally secreted protein, lysozyme, causes the resulting chimeric protein to accumulate in the ER. The lysosomal enzyme cathepsin D with a KDEL sequence added to its carboxy-terminus also accumulates in the ER lumen, but is modified by N-acetylglucosamine phosphotransferase, an enzyme which is resident in the *cis*-Golgi, showing that the cathepsin D had reached the intermediate salvage compartment close to the *cis*-Golgi, but had recycled back to the ER.

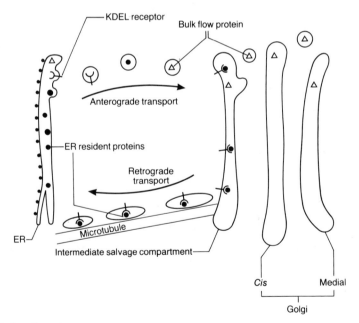

Figure 5.1.   Protein recycling between the ER and Golgi. Proteins are transported in vesicles from the ER to the salvage compartment, where the KDEL receptor carries proteins bound back to the ER by a retrograde transport pathway in which vesicles travel along microtubules.

Two receptors have been identified. A mutant gene, *ERD2*, coding for a 26 kDa receptor protein for the yeast HDEL sequences in mutant strains, allows invertase containing a transplanted HDEL to escape into the medium. A mammalian KDEL receptor (72 kDa) has been identified by making an anti-idiotype antibody; this is prepared by raising an antibody to the binding site of an antibody raised to the targeting peptide (6). A transmembrane glycoprotein *sec20p*, with an HDEL lumenal sequence, identifies the direction the vesicle moves in yeast.

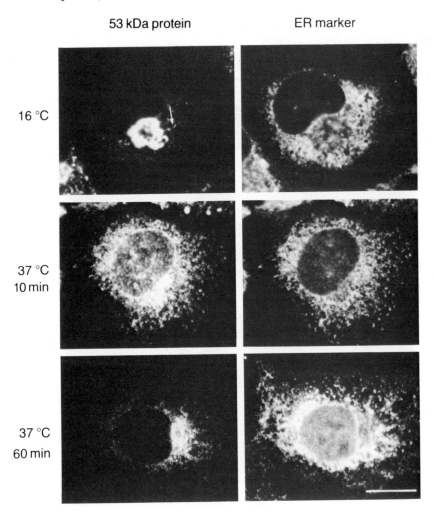

Figure 5.2. Immunofluoresence microscopy of ER/Golgi recycling. In M1 cells at 16°C the 53 kDa marker was localized in a salvage compartment which largely overlaps the Golgi and in a long tubular process extending out of it (marked with an arrow) while the ER marker had a diffuse reticular staining. After 10 minutes at 37°C, the 53 kDa marker adopted the reticular pattern, but relocated around the Golgi after a further 60 minutes. We are grateful to Drs J.Lippincott-Schwartz and R.Klausner for this photograph, courtesy of *Cell*.

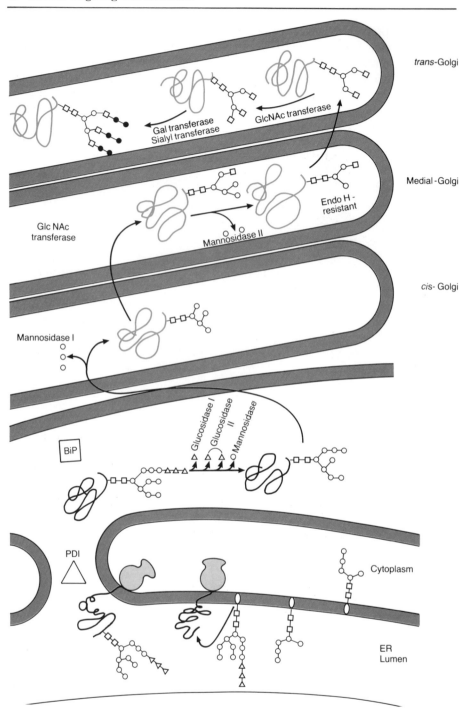

Figure 5.3. Glycosylation in the ER and Golgi. Successive processing of glycoproteins occurs as they move from the ER through successive stacks of the Golgi.

The K(H)DEL receptors bind to proteins containing KDEL or HDEL that have escaped from the ER with higher affinity in the intermediate salvage compartment, and carry the protein back to the ER where they are released (*Figure 5.1*) to be recycled back to the salvage compartment.

An ER retention sequence is contained in the cytoplasmic domain of the integral membrane glycoprotein E19, encoded by the adenovirus, which binds to the Class I major histocompatibility complex in the ER to prevent its expression at the cell surface. This allows the virus to evade immune surveillance as no T-cells are activated. The retention sequences of this and other ER membrane proteins are hydrophilic and possess a lysine residue next-but-one to the carboxy-terminus (*Table 1.1*), and may attach to the microtubule network that is responsible for maintaining the tubular morphology of the ER.

2.3 ER recycling pathway

Proteins are transported to and from the ER and Golgi by vesicular flow. Placing M1 fibroblast cell lines at 16°C inhibits the retrograde pathway from the *cis*-Golgi to the ER, but allows the anterograde transport of proteins outward from the ER to continue. Thus, at lower temperatures, a recycling marker (53 kDa) protein can be seen to accumulate in a salvage compartment close to the Golgi (7) (*Figure 5.2*), and redistributes to the ER only after raising the temperature of the cells to 37°C for 10 minutes. After a longer incubation, the marker is spread over both compartments.

3. Glycosylation

Oligosaccharide moieties containing mannose, N-acetylglucosamine, and glucose are transferred from dolichol precursors into a covalent N-glycosidic linkage to asparagine residues in some nascent chains containing the consensus sequence Asn – X – Thr (or Ser) as they emerge into the ER lumen (8). The carbohydrate moieties are processed further by trimming or addition of further monosaccharide residues either in the ER, or as the glycoproteins pass through successive stacks of the Golgi (*Figure 5.3*). Identification of the extent of carbohydrate modification of a glycoprotein, such as the G-protein encoded by VSV, is a useful experimental tool for monitoring progress through the Golgi (9).

4. Budding and fusion of transport vesicles

Budding from the *trans*-Golgi to produce transport vesicles for the regulated secretory pathway, or the endosomal pathway to lysosomes, occurs at clathrin-coated regions. Vesicles from the ER or Golgi stacks that carry the bulk, unregulated flow of proteins to the cell surface bud off at regions that are coated with a non-clathrin coat material. The drug brefeldin A prevents binding of a 110 kDa protein to form the coat, and causes tubes to form between the Golgi and ER, instead of vesicles. Decoating of clathrin, which must happen before fusion to the acceptor compartment can take place, requires ATP, while decoating of non-clathrin proteins requires GTP hydrolysis; the presence of the non-hydrolysable analogue GTP-γ-S leads to build-up of unfused vesicles.

The biochemistry of the transport of VSV G-glycoprotein through the Golgi has been studied in cultured infected mammalian cells which have been semi-permeabilized by membrane-disrupting agents such as streptolysin or saponin, or by mechanical shear. Vesicle fusion requires a soluble tetrameric protein of 76 kDa, known as NSF on account of its sensitivity to the sulphydryl reagent N-ethyl-maleimide (10). NSF works in conjunction with various attachment proteins known as SNAPs (35 kDa, 36 kDa, and 39 kDa) which form a cytoplasmic complex with NSF (*Figure 5.4*) which is recognized by an acceptor membrane.

A 21 kDa GTPase protein known as rab (or ypt 1 in yeast) is also involved in vesicle fusion, ensuring that transport is unidirectional. The hydrolysis of GTP by rab, attached to the transport vesicle via its carboxy-terminal cysteine residue, occurs when contact is made with an activating protein (GAP) integrated into the acceptor membrane, thus preventing return to the donor compartment (*Figure 5.4*).

A number of yeast secretory (*sec*) temperature-sensitive mutants that do not export from the ER to the Golgi at the restrictive temperature are known (11). These are classified into two groups; *sec12*, *sec13*, *sec16*, and *sec23* which cannot produce vesicles, while *sec17*, *sec18*, and *sec22* mutants cannot incorporate them.

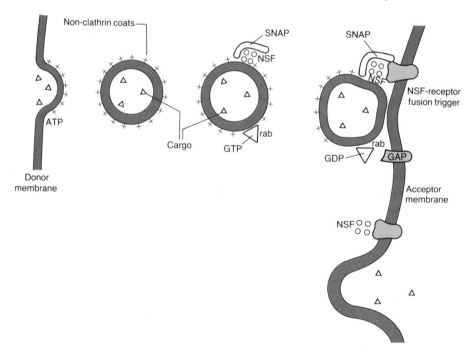

Figure 5.4.   Budding and fusion of vesicles. Role of proteinaceous factors of cargo-carrying vesicles in budding and fusion reactions of vesicles.

Double mutants from the same class produce a lethal phenotype, synthetic lethality (see also Chapter 3, Section 5), suggesting that proteins encoded by the mutant genes interact with each other in such a complex manner that even at the permissive temperature a non-functional complex is produced. *Sec18* encodes for NSF, and *sec17* encodes for one of the SNAPS; the lethality of the *sec17/18* double mutants confirm that these two proteins interact.

5. Sorting at the *trans*-Golgi network

The *trans*-Golgi network is the site of further sorting of proteins between the regulated and constitutive pathways, and the lysosomal pathway (*Figure 1.2*).

5.1 Diversion in to the regulated secretory pathway

Proteins such as the digestive enzymes of the pancreatic acinar cell or polypeptide hormones of the anterior pituitary are concentrated and packaged in a semi-crystalline form, the formation of which is caused by a local drop in pH brought about by a proton-pumping ATPase (12), into secretory granules in the *trans*-Golgi network. The granules are moved to the cell surface along microtubules, and on receipt of a trigger from a ligand–receptor interaction, for example cholecystokinin or acetylcholine at the cell surface, the contents are released into the extracellular fluid by a process of exocytosis when secretory granules fuse with the plasma membrane. Phospholipids and some constitutive proteins of the granules are then retrieved from the membrane back in to the *trans*-Golgi for further use.

Sorting into the regulated pathway is mediated by the receptor-mediated recognition of the aggregates. Matrix granule proteins such as chromogranin A and B, and secretogranin are known to aggregate *in vitro* in the presence of calcium when the pH is lowered (13), but how prohormone and protein precursors become aggregated and concentrated is not known. A signal patch in the aggregate is recognized by a receptor which carries it to a specific portion of the *trans*-Golgi network where granule formation takes place (*Figure 5.5*).

Many secretory proteins and hormones reach the *trans*-Golgi as 'pro' forms which are proteolytically processed during packaging. For example, in the anterior lobe of the pituitary, the 29 kDa precursor of adrenocorticotrophin (ACTH) is cleaved at pairs of basic residues, which are then removed by carboxypeptidase action to release the active hormones ACTH and β-lipotropin. The pro-sequences of some prohormones form part of the signal patch for the regulated pathway.

Small GTP-binding proteins of the *ras*-family are involved in the fusion of secretory granules with the plasma membrane prior to exocytosis. In yeast, in which all secretion is constitutive, fusion requires at least 10 genes. One of these encodes a GTP-binding protein *sec4p*, which attaches to a secretory vesicle *via* covalently-linked fatty-acids. GTP is hydrolysed when the vesicle docks with the plasma membrane.

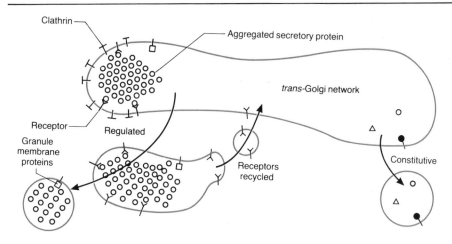

Figure 5.5.   Packaging into secretory granules. Secretory proteins are concentrated at low pH in the *trans*-Golgi network to form aggregates which are recognized by receptors. After condensation of cargo, together with some granule membrane proteins, into secretory granules, the receptor is recycled back to the *trans*-Golgi.

5.2 Formation of mannose-6-phosphate in lysosome precursors

Targeting of lysosomal glycoprotein enzymes begins as they pass through the *cis*-Golgi. Signal patches are recognized by N-acetylglucosamine phospho-transferase which adds 1-phospho-N-acetylglucosamine from the donor UDP-N-acetylglucosamine to several mannose residues on asparagine-linked oligosaccharides. A phosphoglycosidase then removes the terminal N-acetylglucosamine leaving terminal mannose-6-phosphate (man-6-P) markers.

In humans, defects in the transferase cause lysosomal enzyme storage disease (14). In patients with I-cell (inclusion cell) disease, lysosomal enzymes in some cell types fail to receive the man-6-P marker and are secreted, allowing their substrates to accumulate as inclusion bodies. In a related disease, pseudo-Hurler polydystrophy, transferase activity in fibroblasts is reduced by failure to recognize signal patches on the lysosomal pro-enzyme.

5.3 Mannose-6-phosphate receptors

In the *trans*-Golgi, lysosomal enzyme precursors carrying the man-6-P markers are recognized by a man-6-P receptor (15), and clathrin-coated vesicles bearing the complex bud off from a specialized region of the network and deliver their contents to the late endosomal sorting compartment which has a reticular-vesicular structure (*Figure 5.6*) (16). At the low pH of this compartment the complex dissociates, the phosphate group is removed from man-6-P residues and enzyme is transported to the lysosome. The receptor recycles back to the Golgi (*Figure 1.2*).

Two membrane-bound man-6-P receptors are present in most cell types. One, a 300 kDa glycoprotein binds man-6-P in the absence of cations, and contains

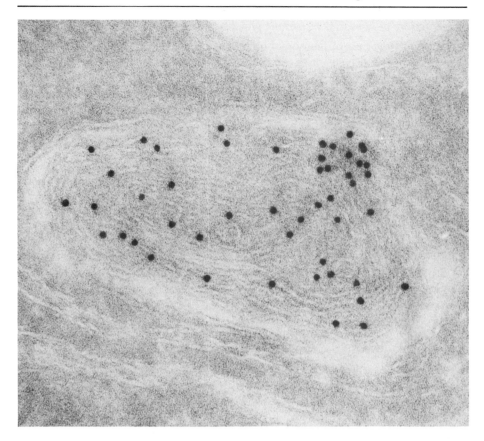

Figure 5.6.   Pre-lysosomal sorting. Antibodies to the 300 kDa man-6-P receptor label a pre-lysosomal late endosomal compartment in the vicinity of the Golgi apparatus. Most of the labelling was located in the worm-like tubules packed to high density inside the lumen of the compartment. The print was provided by Dr Gareth Griffiths of the EMBO Laboratory (× 174 900), courtesy of *Cell*.

a large lumenal domain with 15 sequence repeats of about 145 residues in length. The cation-dependent 45 kDa receptor is a dimer in which each subunit contains a domain homologous to the repeat unit in the 300 kDa receptor.

5.4 Recycling of man-6-P receptors to the cell surface

Man-6-P receptors arrive at the cell surface, either by transport from the late endosome to the early endosome, and then by vesicular transport to the cell surface, or by leakage through the secretory pathway. Thus, proteins containing the man-6-P marker are taken up at the cell surface by endocytosis at coated pits, transferred to the early endosome and then to the lysosomes via the late endosome.

5.5 Targeting of lysosomal membrane proteins

The cytoplasmic tails of lysosomal membrane proteins contain information for their movement to lysosomes. The cytoplasmic tail sequence of a lysosomal protein gp120, for example, on the carboxy-terminal side of the stop-transfer sequence in the membrane is – GRKRSHAGYQTI. Site-specific mutagenic changes of either the Gly(G) or the Tyr(Y) adjacent residues in this sequence yield membrane proteins that are assembled into the plasma membrane rather than in lysosomes. Targeting is thought to be achieved by the way these cytoplasmic tails interact with clathrin and clathrin-binding proteins.

5.6 Targeting to the yeast vacuole

The yeast vacuole contains hydrolytic digestive enzymes such as carboxypeptidase Y (CPY), proteinases A and B, and diaminopeptidase that are analogous to lysosomal enzymes. The sorting mechanisms do not involve man-6-P, and in the case of CPY rely on a hydrophilic targeting sequence – Gln – Arg – Pro – Leu – found in the pro-region, that is a section of the protein found in a precursor sequence which is cleaved off after arrival of the enzyme in the vacuole.

The complexity of the vacuolar targeting process is shown by the fact that over 600 mutants have been isolated (17), falling into 33 complementation groups. Some of the mutations affect the ability of the vacuolar ATPase to maintain the pH at 5.5, while others affect fusion of vesicles with the vacuole which is induced by phosphorylation of vps33, a 691 residue protein, by vps15, a 1409-residue protein. Vps15 is attached to a transport vesicle carrying a lysosomal enzyme by an N-terminal myristoyl group.

6. Endocytosis

6.1 Coated pit formation

Virtually all eukaryotic cells ingest bits of their plasma membrane in the form of small endocytic vesicles, which are later returned to the cell surface. This cycle begins at specialized regions of the membrane known as coated pits. In electron micrographs of the cytoplasmic surfaces of human fibroblast membranes (*Figure 5.7*) (18) coated pits appear as a network of interlocking hexagons and pentagons composed of clathrin, a three-legged (triskelion) protein complex of three 180 kDa and three 33 – 36 kDa molecules. Endocytosis is selective for certain receptors, which after binding to their ligands, cluster at the coated pit (19) (*Figure 5.8*). The role of the coated pit is to control receptor clustering and modulate membrane shape during transformation of the membrane into a vesicle (*Figure 5.9*). Coated pits assemble on top of protein complexes which are seen on the electron microscope as particulate material lying under the lattice network. The complexes, shown in *Figure 5.9*, are bilaterally symmetric structures, and the 100 kDa subunits (adaptins) are thought to provide attachment points for triskelions, while the smaller subunits bind the cytoplasmic domains of receptors.

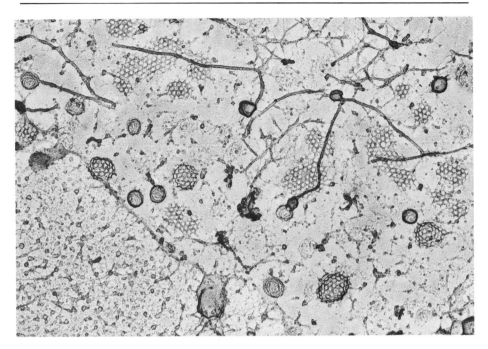

Figure 5.7. Coated pit formation. Shown here is a replica image prepared by a rapid freeze, deep etch method of the cytoplasmic phase of human skin fibroblast membranes, coating with platinum at × 48 720 magnification. Both flat and curved lattices positioned above the membrane surface on top of a protein assembly complex can be seen. Print was provided by Drs David Mahaffey and Richard Anderson at Texas Southwestern Medical Centre (Code 27308).

Adaptor complexes of the plasma membrane contain α- and β-adaptin subunits, while complexes of the *trans*-Golgi network contain γ- and β-subunits. Adaptins have two domains connected by a proline and glycine-rich hinge region.

The low-density lipoprotein receptor contains a sequence Phe – Asp – Asn – Pro – Val – Tyr in its cytoplasmic domain which is responsible for clustering and internalization, and may be the point of attachment to the assembly complex. One in 500 humans has mutant LDL receptors which can bind LDL, but cannot internalize and may be altered in their cytoplasmic tail sequences. These individuals are particularly susceptible to heart attacks. Transferrin, asialoglycoprotein, polyIgA, and Ca^{2+}-man-6-P receptors also contain two close aromatic residues in their cytoplasmic internalization sequences.

6.2 Endosomal sorting

After entering the cell, the endocytosed vesicles lose their clathrin coats with the aid of ATP hydrolysis. The vesicles then fuse with the late endosomes, where the ligand and receptor are sorted into one of four possible pathways (*Figure 5.10*).

1. Class 1 sorting applies to receptors such as the LDL-receptor which are

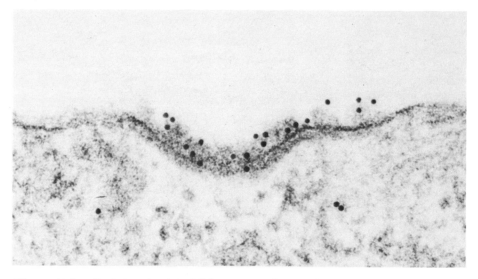

Figure 5.8.   Endocytosis of the LDL-receptor. Coated pits in human skin fibroblasts immunogold labelled with anti-LDL receptor antibody at × 133 350 magnification. The print was provided by Drs Ravindra Pathak and Richard Anderson at Texas Southwestern Medical Centre (Code 26644).

internalized for cell nutrition, the asialoglycoprotein receptor internalized to clear proteins from the circulation, and the insulin receptor, involved in hormone action. The receptors dissociate from the ligands at the low pH of the endosome, the receptor is recycled to the cell surface, and the ligand is transferred to the lysosome.

2. Class 2 sorting applies to the transferrin receptor. In the endosome, iron dissociates, and transferrin together with its receptor recycles to the cell surface, where transferrin dissociates from its receptor.

3. Class 3 sorting applies to growth factor receptors that regulate target cell response by down-regulating receptor concentration; both the ligand and the receptor are transferred to the lysosome.

4. Class 4 sorting applies to transcytosis where a ligand and its receptor after internalization to the endosome is delivered to a different specialized part of the plasma membrane.

Vesicles containing receptors and ligands are delivered to a fusion-competent 'early' endosome, which has a half-life of about eight minutes, during which time recycled components are sent back to the plasma membrane. Recycling requires ATP, NSF, and other cytosolic proteins. During this initial time, LDL accumulates 40-fold, while transferrin accumulates only 3–4-fold. The endosome then becomes resistant to the fusion of further vesicles, and with further lowering of the pH the contents are moved further away from the plasma membrane into the cells's interior, to a late endosome (*Figure 5.10*), from where they move to the lysosomes or the plasma membrane.

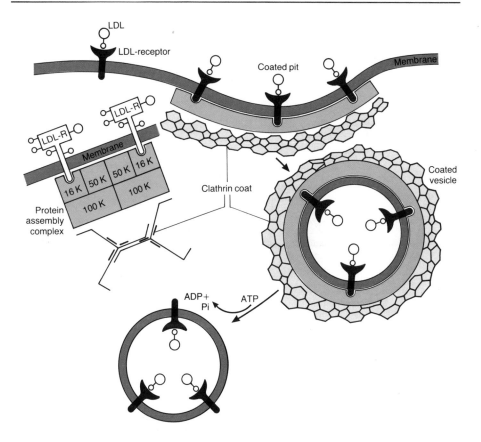

Figure 5.9. Mechanisms of coated pit formation. Clathrin forms where receptors are internalized at coated pits on assembly complexes on the cytoplasmic side of the plasma membrane. Part of the assembly complex, shown in detail, may bind the cytoplasmic tail of the receptor.

7. Sorting in polarized cells

Epithelial cells are organized into sheets by linkage through junctional complexes so that they form selective permeability barriers. Enterocytes, for example, absorb nutrients from their apical microvilli and deliver them to the basolateral surface and into the blood stream. The polarized functions are brought about by an asymmetric distribution of integral membrane proteins ensuring that specific functions are carried out at the different surfaces. Some polarized epithelial cells secrete some of their products at specialized apical surfaces of the plasma membrane. Acinar cells, for example, can secrete digestive enzymes from their apical surface, and components of extracellular matrix from the basolateral surface. These cells segregate plasma membrane components at either surface. Targeting mechanisms are therefore required to sort out which proteins go where.

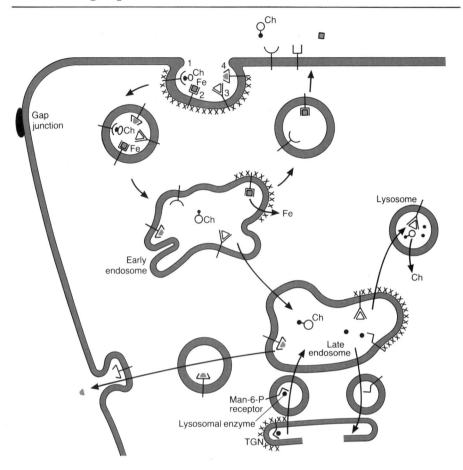

Figure 5.10. Four endocytotic pathways. Shown here are the four different types of sorting occurring after endocytosis of receptors.

In culture, the Madin-Darby canine kidney cell line (MDCK) exhibits all the properties of polarized cells, that is, growing to a tightly sealed mono-layer and exhibiting structural polarity. When infected with influenza virus, the encoded HA protein is delivered specifically to the apical surface, and the VSV-G protein is delivered to the basolateral surface (20). Sorting of these proteins occurs at the *trans*-Golgi, but the nature of the sorting signals is unknown.

In hepatocytes, newly synthesized apical proteins aminopeptidase N and dipeptidylpeptidase IV are targeted first to the basolateral membrane along with basolateral proteins. The apical proteins are then sorted and re-targeted to the basolateral surface at rates that are specific for individual proteins. Hepatocytes constitutively secrete proteins such as albumin and transferrin only from their basolateral surfaces.

7.1 Transcytosis of polymeric IgA

A newborn rat obtains immune protection from its mother's milk by transporting antibodies across the epithelia of the gut. At the low pH of the gut, antibodies bind the polymeric IgA receptor at the apical surface and are internalized *via* coated pits. After passage through endosomes, vesicles fuse with the basolateral domain and antibodies, together with a cleaved portion of the receptor known as a secretory component, are released into the bloodstream. The cytoplasmic tail of the receptor is responsible for targeting (21).

8. Alternative pathways of secretion

A number of proteins including interleukin 1, basic fibroblast growth factor (b-FGF), and lipocortin are not initially synthesized with amino-terminal signal sequences and are secreted from cells by pathways that do not involve the ER (22). The pathway of secretion appears to involve vesicles, and may have evolved from proteins containing sulphydryl groups that would otherwise oxidize in the ER, or meet up with their receptors and lead to cell transformation.

8.1 Secretion of yeast a-factor

Yeast exists in two haploid mating types, each of which produce mating factors a and α. The α-type is made from a precursor with a signal sequence. In contrast, the a-mating factor is a 12-residue peptide generated from a relatively short precursor that lacks a signal sequence and potential glycosylation sites, and is secreted directly from the cytoplasm. The carboxy-terminal cysteine residue is modified with an isoprenoid moiety in thio-ether linkage, and a methyl ester group.

Secretion of the precursor is directly associated with the product of the *STE6* gene, a 1290 residue ATPase with two homologous domains each containing six membrane-spanning domains and one ATP-binding site. The *STE6* gene is homologous to the ATP-driven transport proteins, including the *HylB* gene which is involved in the secretion of haemolysin (see Chapter 2, Section 7), the multi-drug resistance protein (MDR) which pumps drugs out of cells, and leads to the acquisition of drug resistance (23) and CFTR, the protein defective in the disease cystic fibrosis. It is thought that *STE6* forms an ATP-operated channel which opens to allow transport of the mating factor.

9. Further reading

Pfeffer,S.R. and Rothman,J.E. Biosynthetic protein transport and sorting by the endoplasmic reticulum and golgi. (1987) *Ann. Rev. Biochem.*, **56**, 829–52.

Rose,J.K. and Doms,R.W. Regulation of protein export from the endoplasmic reticulum. (1988) *Ann. Rev. Cell Biol.*, **4**, 257–88.

Warren,G. Salvage receptors: two of a kind? (Review). (1990) *Cell*, **62**, 1–2.

Bourne,H.R. Do GTP-ases direct membrane traffic in secretion. (1988) *Cell*, **53**, 669–71.

Dahms,N.M., Lobel,P. and Kornfeld,S. Mannose-6-phosphate receptors and lysosomal enzyme targeting. (1989) *J. Biol. Chem.*, **264**, 12115–18.
Shepherd,V.L. Intracellular pathways and mechanisms of sorting in receptor-mediated endocytosis. *Trends in Pharmacological Sciences*, **10**, 458–62.

10.   References

1. Tartakoff,A.M. and Vassalli,P. (1978) *J. Cell. Biol.*, **79**, 694–707.
2. Lodish,H.H., Kong,N., Snider,M. and Strous,G.J.A.M. (1983) *Nature*, **304**, 80–3.
3. Gething,M.-J., McCannon,K. and Sambrook,J. (1986) *Cell*, **46**, 939–50.
4. Bulleid,N.J. and Freedman,R.B. (1988) *Nature*, **335**, 649–51.
5. Munro,S. and Pelham,H.R.B. (1987) *Cell*, **48**, 899–907.
6. Vaux,D., Tooze,J. and Fuller,S. (1990) *Nature*, **345**, 495–502.
7. Lippincott-Schwartz,J., Donaldson,J.G., Schweizer,A., Berger,E.G., Hauri,H.-P., Yuan,L.C. and Klausner,R.D. (1990) *Cell*, **60**, 821–36.
8. Rothman,J.E. and Lodish,H.F. (1977) *Nature*, **269**, 775–80.
9. Dunphy,W.G. and Rothman,J.E. (1985) *Cell*, **42**, 13–21.
10. Beckers,C.J.M., Block,M.R., Glick,B.S., Rothman,J.E. and Balch,W.E. (1989) *Nature*, **339**, 397–400.
11. Novick,P., Field,C. and Scheckman,R. (1980) *Cell*, **21**, 205–15.
12. Anderson,R.G.W. and Pathak,R.K. (1985) *Cell*, **40**, 635–43.
13. Gerdes,H.-H., Rosa,P., Phillips,E., Baeuerle,P.A., Frank,R., Argos,P. and Huttner,W.B. (1989) *J. Biol. Chem.*, **264**, 12009–15.
14. Kornfeld,S. (1986) *J. Clin. Invest.*, **77**, 1–6.
15. Kaplan,A., Achord,D.T. and Sly,W.S. (1977) *Proc. Natl. Acad. Sci. USA*, **74**, 2026–30.
16. Griffiths,G. (1989) *J. Cell Sci. Suppl.*, **11**, 139–47.
17. Banta,L.M., Robinson,J.S., Klionsky,D.J. and Emr,S.D. *J. Cell Biol.*, **107**, 1369–83.
18. Heuser,J. (1980) *J. Cell Biol.*, **84**, 560–83.
19. Goldstein,J.L., Basu,S.K. and Brown,M.S. (1983) *Methods Enzymol.*, **98**, 241–60.
20. Rodriguez-Boulan,E. and Sabatini,D.D. (1978) *Proc. Natl. Acad. Sci. USA*, **75**, 5071–5.
21. Mostov,K.E., deBruyn,K.A. and Deitcher,D.L. (1986) *Cell*, **47**, 359–64.
22. Rubartelli,A., Cozzolino,F., Talio,M. and Sitia,R. (1990) *EMBO J.*, **9**, 1503–10.
23. McGrath,J.P. and Varshavsky,A. (1989) *Nature*, **340**, 400–4.

6

Import of proteins from the cytosol

This chapter deals with targeting of recently synthesized proteins into subcellular organelles directly from the cytoplasm, that is by pathways that do not tranverse the rough ER.

1. Mitochondria

Proteins are imported into one of four compartments: the matrix, the inner membrane, the intermembrane space (IMS), and the outer membrane. Import processes have been studied *in vitro* by incubating energized mitochondria with newly synthesized precursor proteins or chimeric proteins containing mitochondrial transit sequences fused to cytoplasmic proteins (*Figure 6.1*) in cell lysates (1). Import is post-translational and requires energy. Hydrolysis of nucleotide triphosphates is required, plus a membrane potential across the inner membrane for initial insertion of the transit sequence of those proteins that cross the inner membrane (2).

Many precursors contain targeting sequences of 10–70 amino acid residues, which in many cases are localized at the N-terminus and are cleaved off after import. Quite a number of mitochondrial proteins are synthesized without cleavable transit sequences. The inner membrane ADP/ATP carrier and uncoupling protein for example possess several internal targeting sequences.

1.1 Transit sequences

The essential portion of a transit sequence is frequently shorter than the sequence that is cleaved; for example the first 12 residues of the 25-residue long cleaved transit sequence of cytochrome oxidase IV is sufficient. Typical transit sequences, listed in *Table 6.1*, contain toward their N-termini high numbers of positively charged arginine and lysine residues, hydroxy amino acids, and small groups of hydrophobic residues.

Fusion protein

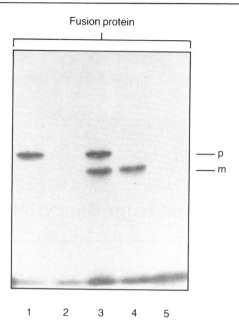

— p
— m

1   2   3   4   5

Figure 6.1.   SDS-PAGE of translated and imported protein. DNA coding 22 amino acids of a mitochondrial transit sequence fused to the coding sequence of cytosolic DHFR placed downstream of a strong promoter was transcribed *in vitro*. After translation in a reticulocyte lysate with radioactive methionine, the labelled fused protein (p) was not processed by de-energized mitochondria (track 1), but was imported and the transit sequence was cleaved to give a smaller form (m) by energized mitochondria (tracks 3 and 4). Import rendered m resistant to added proteinase (track 4), except when mitochondrial membranes were dispersed by detergent (track 5). By kind permission of Professor G.Schatz, courtesy of *FEBS* Letters.

The sequences of transit peptides are not highly conserved, but have been shown by CD spectra that they are capable of adopting α-helical structure in apolar solvents, detergents, or acidic phospholipids such as cardiolipin, which are prominent in mitochondrial membranes. When viewed down the axis of the α-helix, transit peptides have amphiphilic structures with charged residues lying on one side of the helix, and hydrophobic residues lying principally on the other (*Figure 6.2*). As an amphiphilic helix, transit sequences may interact with and disrupt phospholipid bilayers by adopting an orientation in which its basic residues lie on the surface and interact with acidic groups of phospholipids; the non-polar face is buried. Thus, transit sequences may disrupt mitochondrial membranes, facilitating import of the attached protein precursors (3).

1.2 Import mechanisms

The amphiphilic N-terminal targeting sequences of many mitochondrial protein precursors are first recognized at the outer membrane by a receptor protein known as MOM19 (4). A separate receptor, MOM72 (5) binds the internal transit

Table 6.1.   N-Terminal sequences of some mitochondrial precursors

Outer membrane: 70 kDa protein (yeast)

```
+ +    o + +o       o      o o                      +  ++ o    +
MKSFITRNKTAILATVAATGTAIGAYYYYNQLQQQQQRGKKNTINK-
--++
DRKK
```

Intermembrane space: Cytochrome c peroxidase (yeast)

```
+ o o    +     o    +o    + +o      o            o     o o  + +ooo
MTTAVRLLPSLGRTAHKRSLYLFSAAAAAAAAAAATFAYSQSHKRSSS-
o     o             +          o o
SPGGGSNHGWNNWGKAAALAS^T
```

Inner membrane – c side: Cytochrome c_1 (yeast)

```
+  o  o++       +o o+o  oo o      o+o  + o  +   o
MFSNLSKRWAQRTLSKSFYSTATGAASKS^GKLTQKLVTAGVAAAGI-
o  oo           -  -
TASTLLYADAEA^M
```

Inner membrane – intrinsic: Cytochrome oxidase IV (bovine)

```
+    o+  o    ++ ooo    +   + o
MLATRVFSLIGRRAISTSVCVR^AHGSVV
```

Matrix: Malate dehydrogenase (rat)

```
+  o   +        ++o ooo      +
MLSALARPVGAALRRSFSTSAQNN^AKVAVL
```

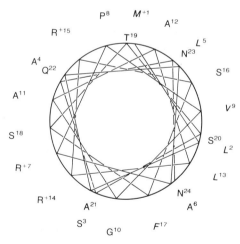

Figure 6.2.   An amphiphilic helix. The positively charged and hydrophobic (bold) residues lie on opposite sides of an α-helix (shown when viewed down the axis) of the transit sequence of malate dehydrogenase.

sequences found in the ATP/ADP carrier. Translocation is known to occur at contact sites between the inner and outer membranes, and accumulated precursors at these sites have been localized by labelling with protein A gold (5). Having captured the precursor proteins from the cytoplasm, it is postulated

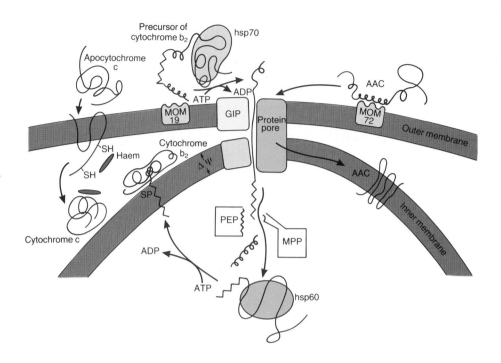

Figure 6.3. Mitochondrial import and sorting mechanisms. The import pathway of cytochrome b_2 is through to the matrix at the contact sites, where the N-terminal transit sequence is removed by the matrix processing peptidase (MPP) in concert with the peptidase enhancing protein (PEP) exposing a second signal sequence. After stabilization with hsp60, the intermediate is transported back to the intermembrane space, where the signal sequence is cleaved off by a peptidase (SP). The ATP/ADP carrier (AAC) is imported via a different receptor through the same general import protein (GIP), and then moved directly to the inner membrane. Apocytochrome c is imported directly across the outer membrane into the IMS, where it folds after addition of haem.

that the receptors deliver them to a general import protein in the outer membrane (GIP) (*Figure 6.3*).

A 42 kDa protein at the import site is cross-linked by a fused protein containing, at the amino-terminus, the transit sequence of yeast cytochrome subunit IV joined to mouse DHFR. The DHFR moiety contains a single cysteine residue at the C-terminus linked to the N-terminus of bovine trypsin inhibitor through a bifunctional cross-linker containing a photoreactive group (*Figure 6.4*) (6). Translocation across the inner membrane depends on the electrical component of the electrochemical potential, and is either through a protein tunnel, of which the 42 kDa protein may be a subunit, or directly though the phospholipid bilayer that has been disrupted by the amphiphilic transit sequence. The positive charge on the outside and negative charge on the matrix side, which is generated by

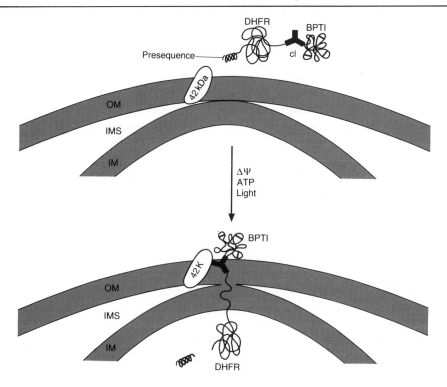

Figure 6.4. Crosslinking from a chimeric protein wedged in the mitochondrial import tunnel. A fused protein construct containing a transit (pre) sequence, DHFR, a photoactivatable cross-linker (cl) then trypsin inhibitor (BPTI) gets stuck after insertion into energized mitochondria, and under UV light cross-links to a 42 kDa protein. OM, outer membrane; IM, inner membrane; IMS, inter membrane space. We are grateful to Dr G.Schatz, courtesy of Macmillan Magazines, for this picture.

the electron-transport chain, may drive the transit peptide through the inner membrane by an electrophoretic effect.

In the matrix, the transit sequences are removed by the divalent cation-dependent processing peptidase (MPP) in concert with an activating protein, protease enhancing protein (PEP). In the matrix, proteins bind to hsp60 which, with the aid of ATP hydrolysis helps them adopt their final conformation (7).

1.3 Further sorting

Proteins such as porin or the mitochondrial 70 kDa protein are recognized by binding sites on the outer membrane, then assembled directly into the outer membrane. Proteins such as cytochrome c_1 of the bc_1 complex which is assembled on the outer surface of the inner membrane, and cytochrome b_2, a soluble IMS protein, have complex presequences (*Table 6.1*). They are first imported into the matrix at contact sites by an N-terminal amphiphilic sequence as described previously. The matrix targeting sequence is then cleaved off to expose an additional stretch of about 20 consecutive hydrophobic residues preceded by some basic residues, similar in structure to the signal sequences that target proteins to the bacterial periplasmic space. The signal sequence directs transport of the intermediate form of the protein back to the intermembrane space, where a second cleavage occurs (*Figure 6.3*). The similarity of this second stage of transport to protein export across the bacterial plasma membrane is evidence that mitochondria evolved from prokaryotic ancestors by endo-symbiosis, and have retained the prokaryotic transport machinery.

1.4 Import of apocytochrome c

Import of cytochrome *c* occurs directly across the outer membrane, does not require an electrical potential, and is not accompanied by proteolytic processing (*Figure 6.3*). Apocytochrome *c* translocates spontaneously through the outer membrane in an unstructured form, then the haem prosthetic group is donated in the IMS from cytochrome *c* lyase by a process that requires NADH plus FMN. The folding that accompanies haem addition is responsible for driving translocation to completion.

2. Chloroplasts

While about 20 per cent of chloroplast proteins are synthesized in the chloroplast stroma on prokaryotic-type ribosomes, the rest are encoded by the plant nuclear genome and are synthesized in the cytoplasm. After import, sorting into one of six different compartments takes place namely; to the outer or inner membranes, the intermembrane space, stroma, thylakoid membrane, and thylakoid lumen (*Figure 6.5*).

Ribulose-1,5-bisphosphate carboxylase (Rubisco), a plentiful stromal enzyme responsible for carbon fixation, consists of eight chloroplast-encoded large subunits and eight small subunits which are imported from the cytoplasm. The initially translated form of the small subunit contains a transit sequence of about 40 residues which is cleaved in the stroma (8) (*Table 6.2*). Fusion proteins containing this transit sequence and a bacterial cytoplasmic protein neomycin phosphotransferase are successfully imported into chloroplasts (9) showing that all the information required for transport resides in the transit sequence. The precursors of ferredoxin, a small protein containing an iron – sulphur complex which resides on the stromal surface of the thylakoid membrane, and a light harvesting chlorophyll a/b binding protein in the thylakoid membrane on the stromal side also contain transit sequences (*Table 6.2*).

2.1 Structures of transit sequences

Transit sequences are 30 – 80 amino acid residues in length. Except for Met – Ala at the amino-terminus, consecutive serines or threonines, and the sequence Leu – Lys in the central portions, these sequences are not conserved. Chloroplast transit peptides contain few acidic amino acids, and are relatively rich in serine,

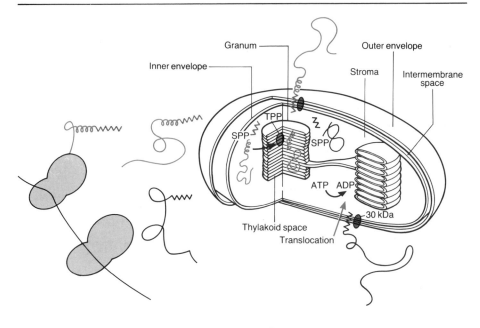

Stromal transit peptide
Thylakoid transit peptide

Figure 6.5. Import pathways into chloroplasts. A stromal protein precursor is imported via a 30 kDa receptor interacting with its transit sequence, which is then cleaved by a stromal peptidase (SPP). In thylakoid proteins, a signal sequence is revealed, which is cleaved by a peptidase (TPP) after movement to the thylakoid.

threonine, and alanine residues. They have less basic residues than mitochondrial transit sequences, and the groups of hydrophobic residues are separated. CD and 2D-NMR show them to be helical in apolar environments, with a kink at the proline residue, but they do not exhibit the pronounced amphiphilicity of mitochondrial transit sequences.

2.2 Import across the envelopes

In *in vitro* systems radiolabelled precursor proteins are imported by purified chloroplasts in the absence of ongoing translation (10). Initial binding to the surface of the envelope requires $50-100~\mu M$ hydrolysable ATP, while the ensuing translocation step requires $250-1000~\mu M$ ATP in the stroma (11). Internal ATP may be generated by light illumination. Unlike mitochondria, chloroplasts do not require an electrochemical potential to import their proteins.

Monovalent F_{ab} fragments of an anti-idiotype antibody (12) raised initially against a synthetic transit peptide bind to a 30 kDa envelope receptor protein and block import of newly synthesized small subunit Rubisco *in vitro*. Immunogold labelling has localized the receptor at contact zones between inner and outer membranes (*Figure 6.6*).

Table 6.2.　Chloroplast transit sequences

Stroma: wheat ss-Rubisco
```
+           oo  o o              +oo            o++o o·  o o     + +
MAPAVMASSATTVAPFQGLKSTAGLPVSRRSGSLGSVSNGGRIRC^M
```

Thylakoid membrane: pea light-harvesting chlorophyll a/b binding protein
```
+     ooooo     oo  +   +     oo        + o
MAASSSSSMALSSPTLAGKQLKLNPSSQELGAARFTM^
```

Stromal surface of thylakoid: white campion ferredoxin
```
+ oo oo o o o        +           oo  o           +     + + o
MASTLSTLSVSASLLPKQQPMVASSLPTNMGQALFGLKAGSRGRVT-
```
```
AM^
```

Thylakoid: white campion plastocyanin
```
+  o  o oo        o      + oooo+   o  +       o + o   o  o +
MATVTSSAAVAIPSFAGLKASSTTRAATVKVAVATPRMS^IKASLKD-
```
```
                              o
VGVVVAATAAAGILAGNAMA^A
```

Thylakoid: pea cytochrome *f*
```
+  o +    o   ++- o +o o        o +    o
MQTRNAFSWIKKEITRSISVLLMIYIITRAPISNA^
```

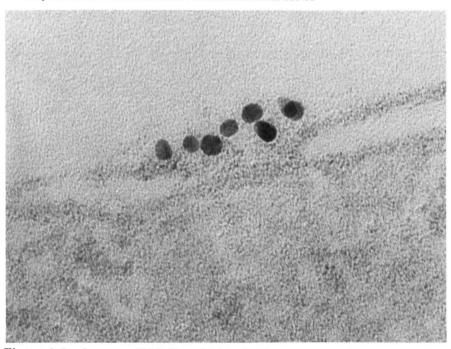

Figure 6.6.　Immunogold labelling of chloroplast import receptor. Electron microscopy of isolated immobilized pea chloroplasts using an anti-idiotype anti-transit peptide antibody. Gold particles are seen at contact sites between inner and outer membranes (× 456 000 magnification). This photograph was provided by Dr D.Pain, courtesy of Macmillan Magazines.

During or shortly after transport into the stroma, the transit sequence is cleaved by a chelator-sensitive stromal peptidase (180 kDa) (13). The role of chaperones in the assembly of Rubisco is described in Chapter 1, Section 8.

2.3 Import into the thylakoid

Transit sequences of thylakoid proteins such as plastocyanin, a copper-containing protein, or the photosystem II water-splitting enzymes have two domains. The amino-terminal domain, which is similar in composition to stromal transit sequence, is followed by a more hydrophobic domain which acts as a thylakoid transfer domain (*Table 6.2*). The hydrophobic domain, as with the mitochondrial intermembrane-space targeting sequence, is similar in structure to bacterial signal peptides in having a positively charged n-region, a hydrophobic h-region, and an alanine residue at the signal peptidase cleavage site.

Import into the thylakoid involves two steps (14). First, the N-terminal 39 residues of the targeting sequences direct the precursors across the envelopes to the stroma, where they are cleaved off by the stromal peptidase. Next, the thylakoid membrane recognizes the hydrophobic domain, and transports the precursor into the lumen, where processing by a protease integrated into the membrane occurs, releasing the mature form of the protein (*Figure 6.5*).

In support of this mechanism is the discovery that cytochrome *f*, which is synthesized in the stroma and then imported into the thylakoid possesses only an amino-terminal hydrophobic signal sequence. In addition, plastocyanin in cyanobacterium, a likely endosymbiotic ancestor of the chloroplast, is made as a precursor with a cleavable hydrophobic signal sequence. These two precursors do not need to cross envelope membranes before reaching their destination.

3. Peroxisomes

Peroxisomes contain a number of enzymes involved in oxidation. These include oxidases which act on various substrates to produce hydrogen peroxide, catalase, which reduces hydrogen peroxide, and enzymes that carry out β-oxidation of fatty acids. Plant glyoxysomes, organelles that contain the five enzymes of the glyoxylate cycle, also contain the oxidases and catalase characteristic of peroxisomes.

Hydrolysis of ATP is required for import, but membrane potential is not required. Peroxisomal proteins are synthesized on free, cytosolic ribosomes and imported across the single peroxisomal membrane post-translationally (15). No modifications of the initially translated form of the protein occurs, and the targeting signal is not cleaved after import. Newly synthesized peroxisomal proteins bind to proteins on the surface of peroxisomes and are translocated in a step which can be prevented by cooling to 0°C.

The targeting sequence lies in the carboxy-terminal sequences. When the gene for firefly luciferase is expressed in mammalian, insect, plant, or yeast cells, luciferase is imported into peroxisomes by virtue of its carboxy-terminal

Table 6.3.   Carboxy-terminal sequences of peroxisomal proteins

Pig D-amino acid oxidase	– Pro – Ser – His – Leu
Rat acylcoenzyme A oxidase	– Gln – Ser – Lys – Leu
Rat enoyl CoA/hydratase-3-hydroxacyl CoA dehydrogenase	– Gly – Ser – Lys – Leu
Human catalase	– Lys – Ser – His – Leu
Candida PMP-20 protein	– Ile – Ala – Lys – Leu
Rat peroxisomal thiolase	– Gly – Ser – Arg – Leu

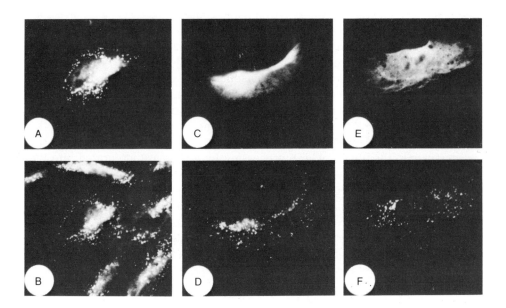

Figure 6.7.   Peroxisomal import of luciferase. A, C, and E show the distribution of luciferase, and B, D, and F show the distribution of catalase in transfected mouse kidney cells in the same field. In A, the localization of luciferase with a C-terminal sequence – Ser – Lys – Leu is the same as catalase (B), in C the C-terminal sequence – Ile – Lys – Leu did not enter into peroxisomes, neither did the C-terminal sequence – Lys – Ser – Lys (E). These photographs are provided by courtesy of Dr Suresh Subramani and the Rockefeller University Press.

sequence, which has been shown to be – Ser – Lys – Leu, and can be visualized by immunofluorescence (*Figure 6.7*) (16). Site-specific mutagenesis and examination of the carboxy-termini of other peroxisomal proteins has shown a requirement for a consensus sequence – Ser(Ala,Cys) – Lys(His,Arg) – Leu (Table 6.3).

A number of genetic disorders involving peroxisomal biogenesis are known in man. Zellweger's Syndrome, which is of autosomal recessive inheritance, involves the mislocation of several peroxisomal enzymes, including catalase, and is likely to be caused by malfunctions in the import machinery.

4. The nucleus

The nucleus is an organelle defined by a boundary known as the nuclear envelope which consists of an inner and outer membrane. The inner membrane is linked to the nuclear lamina and the outer membrane is continuous with the ER. The envelope is perforated by pore complexes, which form channels between the cytoplasm and the nucleus. It is through these pore complexes that mRNA, tRNA, and ribosomal subunits are exported, and proteins are imported. It has been calculated that about 700 ribosomal proteins and 700 histones are imported per minute through one pore.

Released pore complexes appear under the EM as structures with octahedral symmetry (17) (*Figure 6.8*). EM analysis reveals that the pore complex is a

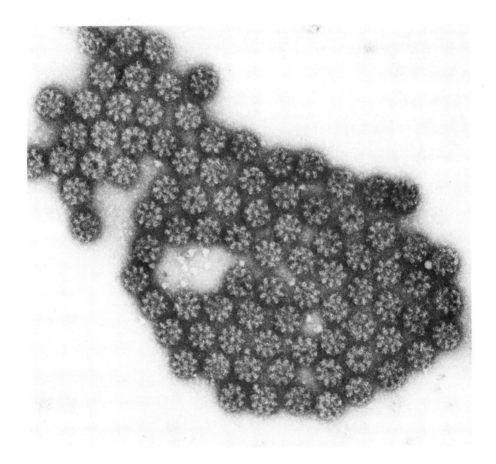

Figure 6.8. Electron micrographs of nuclear pore complexes. The images of pore complexes which have been released from *Xenopus* oocyte nuclear envelopes by detergent treatment, and stained with uranyl acetate (× 86 250 magnification). We are grateful to Dr R.A.Milligan for this photograph, courtesy of the Rockefeller University Press.

symmetrical structure framed by two separated coaxial rings, one attached to the inner membrane on the nuclear side, and one attached to the outer membrane on the cytoplasmic side. The complex has a diameter of about 70 – 100 nm, and an aqueous channel of about 11 nm runs through. Plugs can be seen in the centre of some of the complexes (*Figure 6.8*). The pore complexes contain many different proteins and RNA species. A 210 kDa glycoprotein with large domain in the intermembrane space and a short cytoplasmic domain attaches the complex to the membrane, and glycoproteins containing an unusual O-linked N-acetylglucosamine have been characterized in pore complexes.

Proteins less than 40 kDa diffuse through the pores readily, whereas the transport of larger proteins into the nucleus involves recognition of nuclear targeting sequences, which causes the aqueous channel to widen. Gold-labelled nucleoplasmin, a 165 000 kDa macromolecule, has been visualized passing through the pores into the nucleus (*Figure 6.9*) (18). The nucleoplasmin tail region by itself was found to inhibit transfer, showing that a targeting signal was located there. A positively charged nuclear targeting sequence (*Table 6.4*) was identified by expressing a fused protein in which a transposed sequence from the SV40 T-antigen attached to β-galactosidase caused the enzyme to accumulate in the nucleus (19).

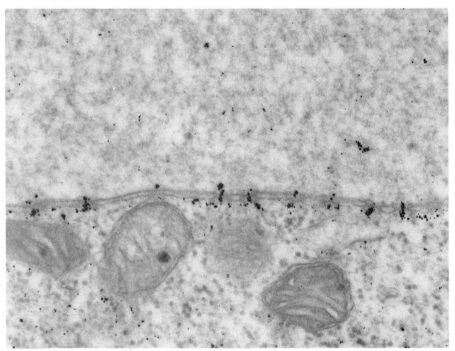

Figure 6.9. Import of gold-labelled nucleoplasmin. Nucleoplasmin-coated gold particles were injected into oocytes and after 15 minutes, the oocytes fixed. Nucleoplasmin is present both in the cytoplasm (bottom of field) and nucleoplasm (top of field), and also extending through pores in the double membrane. (× 60 000 magnification). We are grateful to Dr C.M.Feldherr for this print, courtesy of the Rockefeller University Press.

Table 6.4. Nuclear localization sequences

Adenovirus E1a protein	$\overset{+\,+\;\;\,+}{\text{KRPRP}}$
Polyoma large T antigen	$\overset{+\,+\;\;\,+--}{\text{PKKARED}}$
SV40 large T antigen	$\overset{+++++}{\text{PKKKRKV}}$
Nucleoplasmin	$\overset{++++}{\text{QAKKKKL}}$
P27 post-transcritional regulator (HTLV-1)	$\overset{+\;\;+++\;\;++\;\;\;\;+++}{\text{MPKTRRRPRRSQRKRPPTP}}$

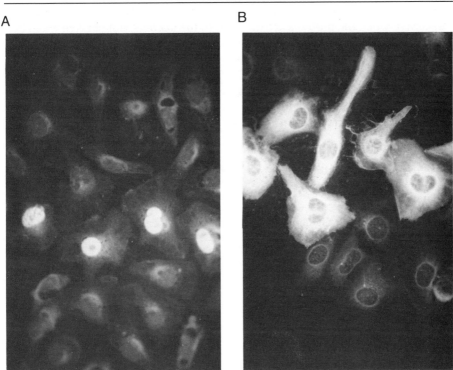

A   B

Figure 6.10. Synthetic peptide nuclear transport signals. Nuclear fluorescence is displayed in cells micro-injected with IgG coupled to a synthetic peptide containing the SV40 targeting sequence (*Table 6.4*) (A), and cytoplasmic fluorescence when the identical peptide, except for replacement of a lysine residue by an asparagine residue, is used (B). We gratefully acknowledge Dr R.E.Lanford for these photographs.

Import of SV40 T-antigen or nucleoplasmin, translated from bacteriophage T7 or SP6 transcripts into isolated nuclei, has been studied *in vitro*. Import, as measured by the degree of protection against added trypsin that has been coupled to agarose, is post-translational and Ca^{2+} and ATP-dependent. Nuclear localization sequences contain groups of basic residues (*Table 6.4*); one of these in a post-transcriptional regulatory protein, targets specifically to the nucleolus.

Non-nuclear proteins coupled chemically to synthetic nuclear targeting peptides and microinjected into cells are imported into the nucleus (*Figure 6.10*) (20). The targeting sequence binds to a soluble 70 kDa receptor protein, identified by chemical cross-linking to a targeting peptide, and by affinity chromatography on a targeting peptide coupled to agarose (21). The receptor may act as a carrier for nuclear-directed proteins between cytoplasm and nucleoplasm and in some way enlarge the pore.

In yeast, mutants have been isolated that exhibit defects in nuclear localization. Chimeric proteins consisting of nuclear localization sequences fused to the mitochondrial proteins cytochrome c_1 and F_1-β-ATPase, expressed in strains which lack mitochondria and are grown on non-fermentable carbon sources, do not survive when the proteins are directed to the nucleus, but survive when the enzymes remain in the cytoplasm. One mutant (*npl1*) codes for a membrane protein which is the same as the protein required for protein transport across the ER (*sec63*) (22) (see Chapter 3).

Thus, transport across the nucleus may be similar in mechanism to transport across other intracellular membranes, but it is only in the nucleus that transport through an aqueous pore can be readily visualized (*Figure 6.9*).

5. Further reading

Mitochondria

Shatz,G. (1987) Signals guiding proteins to their correct locations in mitochondria. *Eur. J. Biochem.*, **165**, 1–6.
Hartl,F.-U., Pfanner,N., Nicholson,D.W. and Neupert,W. (1989) Mitochondrial protein import. (Review). *Biochim. Biophys. Acta.*, **988**, 1–45.

Chloroplasts

Smeekens,S., Weisbeck,P. and Robinson,C. (1990) Protein transport into and within chloroplasts. (Review). *Trends Biochem. Sci.*, **15**, 73–6.

Peroxisomes

Lazarow,P.B. and Fujiki,Y. (1985) Biogenesis of peroxisomes. *Ann. Rev. Cell Biol.*, **1**, 489–530.

Nucleus

Silver,P. and Goodson,H. (1989) Nuclear protein transport. (Review). *Crit. Rev. Biochem. and Mol. Biol.*, **24**, 419–35.

6. References

1. Hurt,E.C., Pesold-Hurt,B. and Schatz,G. (1984) *Febs. Lett.*, **178**, 306–10.
2. Pfanner,N. and Neupert,W. (1985) *EMBO J.*, **4**, 2819–25,
3. Roise,D., Horvath,S.J., Tomich,J.M., Richards,J.H. and Schatz,G. (1986) *EMBO J.*, **5**, 1327–34.

4. Sollner,T., Griffiths,G., Pfaller,R., Pfanner,N. and Neupert,W. (1989) *Cell*, **59**, 1061–70.
5. Schwaiger,M., Herzog,V. and Neupert,W. (1987) *J. Cell Biol.*, **105**, 235–46.
6. Vestweber,D., Baker,A., Brunner,J. and Schatz,G. (1989) *Nature*, **341**, 205–9.
7. Cheng,M.Y., Neupert,W., Martin,J., Hartl,F.U., Hallberg,E.M., Hallberg,R.L., Horwich,A.L., Kalousek,F. and Pollock,R.A. (1989) *Nature*, **337**, 620–5.
8. Schmidt,G.W., Devilliers-Thiery,A., Desruisseaux,H., Blobel,G. and Chua,N.-H. (1979) *J. Cell Biol.*, **83**, 615–22.
9. van den Broeck, Timko,M.P., Kausch,A.P., Cashmore,A.R., Van Montagu,M. and Herrera-Estrella,L. (1985) *Nature*, **313**, 358–63.
10. Chau,N.-H. and Schmidt,G.W. (1978) *Proc. Natl. Acad. Sci. USA*, **75**, 6110–14.
11. Theg,S.M., Bauerle,C., Olsen,L.J., Selman,B.R. and Keegstra,K. (1989) *J. Biol. Chem.*, **264**, 6730–6.
12. Pain,D., Kanwar,Y.S. and Blobel,G. (1988) *Nature*, **331**, 232–7.
13. Robinson,C. and Ellis,R.J. (1984) *Eur. J. Biochem.*, **142**, 337–42.
14. Smeekens,S., Bauerle,C., Hageman,J., Keegstra,K. and Weisbeek,P. (1986) *Cell*, **46**, 365–75.
15. Goldman,B.M. and Blobel,G. (1978) *Proc. Natl. Acad. Sci. USA*, **75**, 5066–70.
16. Gould,S.J., Keller,G-A., Hosken,N., Wilkinson,J. and Subramani,S. (1989) *J. Cell Biol.*, **108**, 1657–64.
17. Unwin,P.N.T. and Milligan,R.A. (1982) *J. Cell Biol.*, **93**, 63–75.
18. Feldherr,C.M., Kallenbach,E. and Schultz,N. (1984) *J. Cell Biol.*, **99**, 2216–22.
19. Kalderon,D., Roberts,B.L., Richardson,W.D. and Smith,A.E. (1984) *Cell*, **39**, 499–509.
20. Lanford,R.E., White,R.G., Dunham,R.G. and Kanda,P. (1988) *Mol. Cell Biol.*, **8**, 2722–9.
21. Adam,S.A., Lobl,T.J., Mitchell,M.A. and Gerace,L. (1989) *Nature*, **337**, 276–9.
22. Sadler,I., Chiang,A., Kurihara,T., Rothblatt,J., Way,J. and Silver,P. *J. Cell Biol.*, **109**, 2665–75.

Appendix

A. Preparation of lysates for cell-free protein translation

The following methods describe preparation of cell-free extracts for the translation of proteins from mRNA. The reagents used for performing the translation assay are described in the references.

A.1 Yeast lysate (Tuite,M.F. and Plesset,J. (1986) *Yeast*, **2**, 35–52.)

Yeast cells were grown to early log phase (OD 600 nm < 1.0), harvested and pre-treated with dithiothreitol prior to spheroplasting with the enzyme Lyticase (Sigma, Poole, UK). The spheroplasts were metabolically regenerated in YM-5 containing 0.4 M magnesium sulphate and homogenized in an all glass Dounce homogenizer. The resulting homogenate was centrifuged at 20 000 g to remove contaminating cellular debris. The supernatant was ultracentrifuged at 100 000 g to remove contaminating cellular organelles and the final supernatant was applied to a Sephadex G-25 column (Pharmacia-LKB, Milton Keynes, UK). The gel filtration separated the yeast lysate from the translation inhibitor and endogenous amino acids. The first peak at OD 260 nm (> 50) was the active translating lysate.

A.2 Reticulocyte lysate (Jackson,R.J. and Hunt,T. (1983) *Methods Enzymol.*, **96**, 50–74.)

Reticulocyte lysate was prepared from lysis of red cells obtained from rabbit blood after anaemia had been induced with acetylphenylhydrazine. Five days after treatment, the rabbits were exsanguinated by cardiac puncture. The red cells were harvested, the packed cell volume being determined, washed in isotonic buffer then lysed with an equal volume of distilled water. The cellular debris was separated by centrifugation at 10 000 g and the supernatant was the active translating lysate.

A.3 Prokaryotic in vitro transcription-translation assay (Zubay,G. (1973) *Annu. Rev. Genet.*, **7**, 267–287.)

The *E. coli* strain was grown to an OD at 450 nm of 1.5–2.0, harvested and stored at −70°C for up to 48 h, thawed and treated with a buffer containing 2-mercaptoethanol, followed by homogenization in a French press. The homogenate was centrifuged at 30 000 g, and the supernatant was incubated with a mixture of phosphoenol pyruvate and amino acids at 37°C and gently agitated.

The mixture was dialyzed against 'S30 buffer' overnight at 4°C, then centrifuged at 4000 g. The final supernatant was the active extract for transcription-translation.

B. Preparation of Microsomes

B.1 Yeast microsomes (Rothblatt,J.A. and Meyer,D.I. (1986) *Cell*, **44**, 619–628.)

Yeast cells were grown to an OD at 600 nm of 2.0–3.0, then spheroplasted and homogenized by the same method used for preparing yeast lysate. The homogenate was layered on to a 0.8 M sucrose cushion and centrifuged at 20 000 g at 4°C to remove cellular debris. The supernatant was ultracentrifuged at 100 000 g. The pellet contained the active microsomes which were taken up in a 'storage buffer' at a concentration of OD at 260 nm of 40.

B.2 Pancreatic microsomes (Kaderbhai,M. and Austen,B.M. (1984) *Biochem J.*, **217**, 145–157.)

The dissected pancreas was suspended in a solution of isotonic sucrose and subjected to two homogenizations, first in a tissue press, then in a motor driven Potter-Elvehjem homogenizer at 4°C. The resultant fine homogenate was centrifuged at 12 000 g to generate a post-mitochondrial supernatant. This was ultra-centrifuged at 100 000 g on a sucrose density gradient between 1.65 M and 2.1 M sucrose. The microsomes located at the interface were diluted, centrifuged and taken up in storage buffer at an OD at 260 nm of 40.

B.3 Prokaryotic inverted vesicles (Chang,C.N., Model,P. and Blobel,G. (1979) *Proc. Natl. Acad. Sci. USA*, **76**, 1251–1255.)

E. coli cells were grown in broth to an OD at 600 nm of 0.5, harvested and resuspended in 10 mM sodium phosphate buffer at pH 7.2. The suspension was sonicated in ten blasts of 30 sec duration, then centrifuged to remove the cellular debris. The supernatant was loaded on a two-step sucrose gradient of 20 and 50 percent (w/v) and ultracentrifuged at 100 000 g. The pellet was subjected to three more ultracentrifugation steps on sucrose gradients. The final pellet was the inverted vesicles which were resuspended in the storage buffer at an OD at 260 nm of 50.

Glossary

Anti-idiotype antibody: an antibody which recognizes the antigen-binding site on a second antibody, and may therefore itself bind to a receptor protein for that antigen.

Assembly: the steps through which a protein takes up its final or mature conformation in which the protein may be inserted into a membrane, or complexed together with other subunits to form an oligomer.

Autoradiography: the detection of radioactivity in a medium by the darkening of a photographic film placed in contact with the medium.

CD: a spectrophotometric technique whereby the secondary structure of a protein or peptide is determined by the way it rotates plane-polarized light of far-UV wavelength.

cDNA: a DNA copy of mRNA.

Chimeric protein: a protein which contains parts of the sequences of other proteins within one polypeptide.

Expression: transcription and translation to produce a protein from coding DNA, often foreign DNA, in a living organism.

Homology: when the sequence of amino acids in protein is genetically related to the sequence in another protein.

Immuno-electron microscopy: view of an antigen in a cell under the electron microscope in which the antibody combines with an electron-dense particle such as gold.

Infra-red spectroscopy: the frequency of absorption of radiation in the infra-red region by peptide bonds. The amide I and amide II bands in peptides or proteins are characteristic of either α-helix, β-sheet or random coil formation.

Integration or Insertion: the process of putting part of a protein's structure across a membrane so it becomes integrated into the phospholipid bilayer.

Lateral pressure: the pressure exerted on a monolayer of phospholipids and/or peptides on an aqueous surface in the plane of the surface.

Leader peptidase: an endoproteinase that cleaves off signal sequences in the bacterial periplasmic space.

Microsomes: a fraction of a tissue homogenate that requires high force (about 100 000 g) to pellet in a centrifuge; often enriched in membranes vesiculated from the rough ER.

Nascent polypeptide: a polypeptide that is in the course of being elongated from a ribosome.

Oligomers: proteins made up of a number of subunits of either identical or different polypeptides.

Retention sequence: a sequence whose overall effect is to retain a protein in a compartment against the unrestricted flow out of it.

SDS-PAGE: electrophoresis of proteins coated with sodium dodecyl sulphate in a gel of polymerized acrylamide; this separates them according to their molecular size.

Site-specific mutations: a way of changing a predetermined site in a protein by changing the corresponding coding triplets in the DNA.

Signal peptidase: an endoproteinase that cleaves off signal sequences in the endoplasmic lumen or mitochondrial inner membrane space.

Signal peptide: a sequence in a protein containing several consecutive hydrophobic residues that bring about translocation across the ER, bacterial inner membrane, thylakoid membrane, or outwards across the mitochondrial inner membrane.

Temperature-sensitive mutations: mutant strains in which the gene product behaves abnormally at one temperature, the non-permissive temperature, and normally at another.

Topology: the topography of a membrane protein is its three-dimensional structure with respect to the membrane. It includes a description of which parts of the protein lie within the bilayer and which parts of the membrane lie on the lumenal or the cytoplasmic side.

Transfection: the introduction of foreign DNA into a living cell.

Transit sequence: an amphiphilic or hydrophilic sequence that brings about translocation into the mitochondrial matrix or chloroplast stroma.

Translation: the production of protein from mRNA by machinery, involving a ribosome in which the sequence of amino acids is dictated by a certain reading frame of triplet codons of bases on the mRNA.

Translocation: movement by a protein from one cellular compartment to another; the process involves crossing a membrane.

Triskelion: a three-legged structure assembled from protein subunits.

2D-NMR: a sophisticated technique in which the position of protons in a three-dimensional space can be determined by measuring the interactions of their magnetic dipoles.

Vesicle: a small, spherical compartment bounded by a phospholipid bilayer containing encapsulated protein. During preparation, some vesicles are inverted so that the outside surface of the original membrane becomes the inside of the vesicle.

Index